LONDON MATHEMATICAL SOCIETY LECTU...

Managing Editor: Professor N.J. Hitchin, Mathematical Institute,
University of Oxford, 24–29 St Giles, Oxford OX1 3LB, United Kingdom

The titles below are available from booksellers, or, in case of difficulty...

London Mathematical Society Lecture Note Series. 289

Aspects of Sobolev-Type Inequalities

Laurent Saloff-Coste
Cornell University

PUBLISHED BY THE PRESS SYNDICATE OF THE UNIVERSITY OF CAMBRIDGE
The Pitt Building, Trumpington Street, Cambridge, United Kingdom

CAMBRIDGE UNIVERSITY PRESS
The Edinburgh Building, Cambridge, CB2 2RU, UK
40 West 20th Street, New York, NY 10011–4211, USA
477 Williamstown Road, Port Melborne, VIC 3207, Australia
Ruiz de Alarcón 13, 28014 Madrid, Spain
Dock House, The Waterfront, Cape Town 8001, South Africa

http://www.cambridge.org

First published 2002

Printed in the United Kingdom at the University Press, Cambridge

A catalogue record for this book is available from the British Library

ISBN 0 521 00607 4 paperback

Contents

Preface

These notes originated from a graduate course given at Cornell University during the fall of 1998. One of the aims of the course was to present Sobolev inequalities and some of their applications in the context of analysis on manifolds —including Harnack inequalities and heat kernel estimates— to an audience not necessarily very familiar with analysis in general and Sobolev inequalities in particular. The first part (Chapters 1 2) introduces the reader to Sobolev inequalities in $\mathbb{R}^n$. An important application, Moser's proof of the elliptic Harnack inequality for uniformly elliptic divergence form second order differential operators, is treated in detail. In the second part (Chapters 3–4), Sobolev inequalities on complete non-compact Riemannian manifolds are discussed: What is their meaning and when do they hold true? How does one prove them? This discussion is illustrated by the treatment of some explicit examples. In the third and last part, Chapter 5, families of local Sobolev and Poincaré inequalities are introduced. These turn out to be crucial for taking full advantage of Sobolev inequality techniques on Riemannian manifolds. For instance, complete Riemannian manifolds satisfying a scale-invariant parabolic Harnack inequality are characterized in terms of Poincaré inequalities and volume growth. These notes give the first detailed exposition of this fundamental result.

We warn the reader that no effort has been made to include a comprehensive bibliography. Many important papers related to the topics presented in these notes are not mentioned. Actually, the literature on Sobolev inequalities is so vast that it would certainly be difficult to list it all. A few of the classical books on the subject have been listed here.

Concerning Riemannian geometry, the books [5, 29] and [12, 13] are very useful references and contain some material related to the present text. There is some overlapping between these notes and the monographs [39, 40], but it may be less than one would think in view of the titles. In particular, the applications presented here and in [39, 40] are different.

Some of the techniques from functional analysis used here are developed in greater generality in [21, 72, 87]. Of these three books, the closest in spirit to these notes might be [21], although there is very little direct overlapping and the two complement each other. Grigor'yan's survey article [34] is a wonderful source of information for many related topics not treated in this monograph.

It is a pleasure to acknowledge the influence, direct or otherwise, that many colleagues and friends had on the writing of this text. Thanks to A. Ancona, D. Bakry, A. Bendikov, T. Coulhon, P. Diaconis, A. Grigor'yan, L. Gross, W. Hebisch, A. Hulanicki, M. Ledoux, N. Lohoué, M. Solomyak, D. Stroock and N. Varopoulos. Thanks to the students and colleagues at Cornell who attended the class on which these notes are based. They helped me to try to stay honest. Finally, I would like to thank the various institutions whose support over the years has made the writing of this book possible. They are, in no particular order, Le Centre National de la Recherche Scientifique, l'Université Paul Sabatier in Toulouse, France, the National Science Foundation (grant DMS-9802855), and Cornell University.

Introduction

This introduction describes some of the main ideas, problems and techniques presented in this monograph.

Chapter 1 gives a brief but more or less self-contained account of Sobolev inequalities in $\mathbb{R}^n$. The Sobolev inequality in $\mathbb{R}^n$ asserts that

$$\left(\int_{\mathbb{R}^n} |f(x)|^{np/(n-p)} dx \right)^{np/(n-p)} \leq C(n,p) \left(\int_{\mathbb{R}^n} |\nabla f(x)|^p dx \right)^{1/p},$$

that is,

$$\|f\|_q \leq C(n,p)\|\nabla f\|_p, \quad q = np/(n-p),$$

for all smooth functions f with compact support and each $1 \leq p < n$. When $p > n$, the Hölder continuity estimate

$$\sup_{x,y \in \mathbb{R}^n} \left\{ \frac{|f(x) - f(y)|}{|x - y|^{1-n/p}} \right\} \leq C\|\nabla f\|_p$$

holds instead. We discuss a number of different proofs of Sobolev inequalities in $\mathbb{R}^n$. Each yields a different and useful point of view on the meaning of Sobolev inequalities. Of course, this material is covered in greater detail in a number of books and monographs including [1, 30, 60, 79]. The important topic of Sobolev inequalities in subdomains of $\mathbb{R}^n$ (see, e.g., [61]) is not treated here.

The theory of partial differential equations provides a host of important applications of Sobolev inequalities. Consider for instance the equation

$$\sum_{i,j=1}^{n} \partial_i a_{i,j}(x) \partial_j u(x) = 0$$

where the coefficients $a_{i,j}$ are real measurable functions such that

$$\|a_{i,j}\|_\infty \leq C_1$$

and

$$\forall\, x \in \mathbb{R}^n, \forall\, \xi \in \mathbb{R}^n, \quad \sum_{i,j=1}^{n} a_{i,j}(x)\xi_i\xi_j \geq c_1 \sum_{1}^{n} \xi_i^2.$$

That is, consider a divergence form, uniformly elliptic equation in $\mathbb{R}^n$. Moser's elliptic Harnack inequality [30, 63] states that any positive weak solution u of this equation in an Euclidean ball B satisfies

$$\sup_{\frac{1}{2}B}\{u\} \le C \inf_{\frac{1}{2}B}\{u\}$$

where C depends neither on u nor on B but only on the constants C_1, c_1 above and the dimension n. Moser's proof, presented in Chapter 2, is a striking application of Sobolev inequalities. It also serves as an introduction to our later treatment of parabolic Harnack inequalities on manifolds.

In Chapter 3, Sobolev inequalities are discussed in the context of Riemannian manifolds. A number of related functional inequalities are introduced and relations between these inequalities are established. One of the most basic facts is that any Sobolev inequality implies a lower bound on the volume growth of the geodesic balls. In particular, the inequality

$$\forall f \in \mathcal{C}_0^\infty(M), \quad \|f\|_q \le C\|\nabla f\|_p$$

for some fixed $q > p \ge 1$, implies that the volume of any ball of radius r must be bounded below by a constant times r^ν with ν related to p, q by $1/\nu = 1/p - 1/q$.

A more technical but very important fact is the equivalence between strong forms and weak forms of Sobolev inequalities. An example of this phenomenon is that it is enough to have the weak Sobolev inequality

$$\forall f \in \mathcal{C}_0^\infty(M), \quad \sup_{s>0}\left\{ s\mu\left(\{x : |f(x)| > s\}\right)^{1/q}\right\} \le C\|\nabla f\|_p$$

with $1 \le p < q$ to conclude that the strong inequality $\|f\|_q \le C\|\nabla f\|_p$ holds (with different constants C). Another example is the equivalence between the Nash inequality

$$\forall f \in \mathcal{C}_0^\infty(M), \quad \|f\|_2^{(1+2/\nu)} \le C\|\nabla f\|_2 \|f\|_1^{2/\nu}$$

and the Sobolev inequality

$$\forall f \in \mathcal{C}_0^\infty(M), \quad \|f\|_{2\nu/(\nu-2)} \le C\|\nabla f\|_2$$

when $\nu > 2$ (again with different C's). The Nash inequality is (a priori) weaker in the sense that it is easily deduced from the Sobolev inequality above and Hölder's inequality. Chapter 3 gives a rather complete treatment of this phenomenon using elementary and unified arguments taken from [6]. Related results and interesting developments concerning Sobolev spaces on metric spaces can be found in [38].

The equivalence between weak and strong forms of Sobolev-type inequalities turns out to be extremely useful when it comes to *prove* that a certain manifold satisfies a Sobolev inequality. This is illustrated in the last section of Chapter 3 where some fundamental examples are treated. A basic tool used here is the notion of pseudo-Poincaré inequality. Given a smooth function f, let $f_r(x)$ denote the mean of f over the ball of center x and radius r. One says that M satisfies an L^p-pseudo-Poincaré inequality if, for all $f \in \mathcal{C}_0^\infty(M)$ and all $r > 0$,

$$\|f - f_r\|_p \le C\,r\,\|\nabla f\|_p.$$

For manifolds satisfying a pseudo-Poincaré inequality, Sobolev inequalities can be deduced from a simple lower bound on the volume growth. This is more precisely stated in the following theorem.

Theorem *Let M be a complete Riemannian manifold. Fix p, ν with $1 \le p < \nu$ and assume that M satisfies an L^p-pseudo-Poincaré inequality. Then the Sobolev inequality*

$$\forall f \in \mathcal{C}_0^\infty(M), \quad \|f\|_{\nu p/(\nu-p)} \le C\|\nabla f\|_p$$

holds true if and only if any ball B of radius $r > 0$ has volume bounded below by $\mu(B) \ge cr^\nu$.

The idea behind this theorem first appeared rather implicitly in [72] in the setting of Lie groups. It was later developed in [6, 19, 74] and other works. To illustrate this result, we treat in detail the case of unimodular Lie groups equipped with a left-invariant Riemannian metric as well as manifolds with non-negative Ricci curvature and maximal volume growth. The L^p-pseudo-Poincaré inequality should be compared with the more classical L^p-Poincaré inequality

$$\forall f \in \mathcal{C}^\infty(B), \quad \left(\int_B |f(y) - f_B|^p dy\right)^{1/p} \le Cr\left(\int_B |\nabla f(y)|^p dy\right)^{1/p}$$

where $B = B(x, r)$ denotes a geodesic ball of radius r and $f_B = f_r(x)$ is the mean of f over B. This last inequality may or may not hold on M, uniformly over all balls $B = B(x, r)$, $x \in M$, $r > 0$. The pseudo-Poincaré inequality may hold for all $r > 0$ in cases where the Poincaré inequality does not (for instance on unimodular Lie groups having exponential volume growth).

Chapter 4 develops two different but related applications of Sobolev-type inequalities. These two applications have been chosen for their importance and their simplicity.

First, we show that Nash inequality is equivalent to a uniform heat kernel upper bound of the form

$$\sup_{x,y \in M} h(t, x, y) \le Ct^{-\nu/2}$$

where $h(t, x, y)$ denotes the fundamental solution of the heat equation

$$(\partial_t + \Delta)u = 0$$

on $(0, \infty) \times M$, with $\Delta = -\text{div} \circ \nabla$. In particular, under a Nash inequality, the heat diffusion semigroup $(H_t)_{t>0}$ is ultracontractive (i.e., sends L^1 to L^∞). This has been developed in the last fifteen years into a powerful machinery which produces Gaussian heat kernel upper bounds. Although this circle of ideas has its roots in Nash's 1958 paper [67], it was only after 1980 that the full strength and the scope of this technique was identified. The books [21, 72, 87] contain different accounts of this topic, various applications and further developments. Here, under the basic hypothesis that

$$\forall t > 0, \quad \sup_{x,y \in M} h(t, x, y) \leq Ct^{-\nu/2},$$

we prove that the heat kernel satisfies the Gaussian upper bound

$$h(t, x, y) \leq C_1 t^{-\nu/2}(1 + d^2/t)^{\nu/2} e^{-d^2/4t}$$

where $d = d(x, y)$ is the Riemannian distance between x and y. Our proof is somewhat different from those found in the literature. It is adapted from [41] and uses complex interpolation as a main technical tool (and, ironically, no Sobolev-type inequality).

The second topic treated in Chapter 4 is a spectral inequality known as the Rozenblum–Lieb–Cwikel estimate. This inequality was first proved in $\mathbb{R}^n$ by Rozenblum in 1972. It asserts that the number of negative eigenvalues of the Schrödinger operator $\Delta - V$ is bounded above by $C(\nu)\|V_+\|_{\nu/2}^{\nu/2}$ as soon as the manifold M satisfies the Sobolev inequality

$$\|f\|_{2\nu/(\nu-2)} \leq C\|\nabla f\|_2.$$

The proof presented here is due to P. Li and S-T. Yau, [55]. A central part of this proof is very close in spirit to Nash's ideas concerning ultracontractivity. It illustrates well what can be done by a skillful use of Sobolev inequality and basic functional analysis.

Despite important examples such as $\mathbb{R}^n$ and hyperbolic spaces, many Riemannian manifolds fail to satisfy a global Sobolev inequality of the form

$$\forall f \in \mathcal{C}_0^\infty(M), \quad \|f\|_{2\nu/(\nu-2)} \leq C\|\nabla f\|_2$$

for some $\nu > 2$. For one thing, such an inequality implies that the volume of any ball of radius r is at least cr^ν for all $r > 0$, ruling out many simple interesting manifolds such as $\mathbb{S}^m \times \mathbb{R}^k$ (the product of an m-sphere by a k-dimensional Euclidean space). More generally, such a global Sobolev inequality requires too much "uniformity" of the Riemannian manifold M. Fortunately, there is a way to cope partially with this difficulty. The idea is to use *families of local Sobolev inequalities* instead of one global Sobolev inequality. For any ball $B = B(x, r)$ on a complete Riemannian manifold, one can find a constant $C(B)$ such that, for any smooth function f *with compact support in B*,

$$\left(\int_B |f|^q d\mu \right)^{2/q} \leq \frac{C(B)r^2}{\mu(B)^{2/\nu}} \int_B \left(|\nabla f|^2 + r^{-2}|f|^2 \right) d\mu$$

where $q, \nu > 2$ are some fixed constants related by $1/q = 1/2 - 1/\nu$. A lot of information is encoded in the behavior of the function $B \mapsto C(B)$. The simplest and perhaps most interesting case is when this function is bounded, that is, $\sup_B C(B) = C < \infty$. This can happen in cases where the global Sobolev inequality

$$\left(\int_M |f|^q d\mu \right)^{2/q} \leq C \int_M \left(|\nabla f|^2 \right) d\mu$$

does *not* hold. For instance, the manifold $\mathbb{S}^m \times \mathbb{R}^k$, $m+k > 2$ does not satisfy any global Sobolev inequality (assuming $m \neq 0$) but satisfies a family of local Sobolev inequalities with $\nu = m + k$, $q = 2\nu/(\nu - 2)$ and $\sup_B C(B) = C < \infty$. In the other direction, the hyperbolic space of dimension n satisfies the same global Sobolev inequality as $\mathbb{R}^n$ but does not have $\sup_B C(B) < \infty$. In fact, as far as many applications are concerned (e.g., heat kernel bounds), a family of local Sobolev inequalities with $\sup_B C(B) < \infty$ contains more useful information than a global Sobolev inequality.

Chapter 5 develops these ideas and culminates with a complete proof of the following theorem, where $V(x, r)$ denotes the volume of the ball of center x and radius r, and d is the Riemannian distance. For any $x \in M$ and $s, r > 0$, let $Q = Q(x, s, r)$ be the time–space cylinder

$$Q(x, s, t) = (s - r^2, s) \times B(x, r).$$

Let Q_+, Q_- be respectively the upper and lower subcylinders

$$\begin{aligned} Q_+ &= (s - (1/4)r^2, s) \times B(x, (1/2)r) \\ Q_- &= (s - (3/4)r^2, s - (1/2)r^2) \times B(x, (1/2)r). \end{aligned}$$

We say that M satisfies the scale-invariant parabolic Harnack principle if there exists a constant C such that for any $x \in M$ and $s, r > 0$, and any positive solution u of $(\partial_t + \Delta)u = 0$ in $Q = Q(x, s, r)$, we have

$$\sup_{Q_-} \{u\} \leq C \inf_{Q_+} \{u\}.$$

Theorem *A complete Riemannian manifold M satisfies the scale-invariant parabolic Harnack principle if and only if M satisfies the doubling property*

$$\forall x \in M, \ \forall r > 0, \quad V(x, 2r) \leq D_0 V(x, r)$$

and the scale-invariant Poincaré inequality

$$\forall\, B = B(x,r), \quad \int_B |f - f_B|^2 d\mu \leq P_0 r^2 \int_B |\nabla f|^2 d\mu$$

where f_B denotes the mean of $f \in \mathcal{C}^\infty(B)$ over the ball B.

In fact, the equivalent properties above are also equivalent to the fact that the heat kernel $h(t, x, y)$ satisfies the two-sided Gaussian estimate

$$\forall\, t > 0, \;\; \forall\, x, y \in M, \;\; \frac{c_1 e^{-C_1 d(x,y)^2/t}}{V(x, \sqrt{t})} \leq h(t, x, y) \leq \frac{C_2 e^{-c_2 d(x,y)^2/t}}{V(x, \sqrt{t})}.$$

Such a two-sided heat kernel bound was first derived for uniformly elliptic divergence form second order differential operators in $\mathbb{R}^n$ by Aronson [3].

These results are taken from [32, 74] (a more complete discussion is given at the beginning of Section 5.5). The equivalence between the parabolic Harnack inequality on the one hand and the (more geometric) doubling property and Poincaré inequality on the other hand is a very useful tool. Both directions of this equivalence are interesting and this illustrated by a few simple examples. For instance, it follows from the theorem above that the parabolic Harnack principle is stable under quasi-isometries.

Chapter 1

Sobolev inequalities in $\mathbb{R}^n$

1.1   Sobolev inequalities

1.1.1   Introduction

How can one control the size of a function in terms of the size of its gradient? The well-known Sobolev inequalities answer precisely this question in multidimensional Euclidean spaces. On the real line, the answer is given by a simple yet extremely useful calculus inequality. Namely, for any smooth function f with compact support on the line,

$$|f(t)| \leq \frac{1}{2} \int_{-\infty}^{+\infty} |f'(s)| ds. \tag{1.1.1}$$

The factor $1/2$ in this inequality comes from the fact that f vanishes at both $+\infty$ and $-\infty$. In this respect, note that if f is smooth but no other restriction is imposed the inequality above may fail.

It is natural to wonder if there is such an inequality for smooth compactly supported functions in higher-dimensional Euclidean spaces. More precisely, for each integer n, can one find $p, q > 0$ and $C > 0$ such that

$$\forall f \in \mathcal{C}_0^\infty(\mathbb{R}^n), \quad \|f\|_q \leq C \|\nabla f\|_p? \tag{1.1.2}$$

Here and in the sequel $\mathcal{C}_0^\infty(\mathbb{R}^n)$ denotes the set of all smooth compactly supported functions in $\mathbb{R}^n$. For $f \in \mathcal{C}_0^\infty(\mathbb{R}^n)$, we set

$$\|f\|_q = \left(\int_{\mathbb{R}^n} |f(x)|^q dx \right)^{1/q}, \quad \|f\|_\infty = \sup_{\mathbb{R}^n} \{|f|\}$$

and

$$\|\nabla f\|_p = \left(\int_{\mathbb{R}^n} |\nabla f(x)|^p dx \right)^{1/p}$$

where $\nabla f = (\partial_1 f, \ldots, \partial_n f)$ is the gradient of f and $|\nabla f| = \sqrt{\sum_1^n |\partial_i f|^2}$ is the Euclidean length of the gradient. In $\mathbb{R}^n$, we denote by $\mu_n = \mu$ the

7

Lebesgue measure and by μ_{n-1} the volume measure on smooth hypersurfaces of dimension $n-1$. When using coordinates $x = (x_1, \ldots, x_n)$, we also write

$$d\mu(x) = dx = dx_1 \ldots dx_n.$$

This question was first addressed in this form by Sobolev in [78] which appeared in Russian in 1938. Fixing a function $f \in \mathcal{C}_0^\infty(\mathbb{R}^n)$ and replacing $x \mapsto f(x)$ by $x \mapsto f(tx)$, $t > 0$, in (1.1.2) yields

$$t^{-n/q}\|f\|_q \leq C\, t^{1-n/p}\|\nabla f\|_p.$$

Letting t tend to zero and to infinity shows that (1.1.2) can only be satisfied if the exponents of t on both sides of the inequality above are the same. That is, (1.1.2) can only be satisfied if

$$\frac{1}{q} = \frac{1}{p} - \frac{1}{n}, \text{ i.e., } q = \frac{np}{n-p}. \tag{1.1.3}$$

For instance, in $\mathbb{R}^2$, this says that one might possibly have

$$\forall\, f \in \mathcal{C}_0^\infty(\mathbb{R}^2), \quad \|f\|_\infty \leq \int_{\mathbb{R}^2} |\nabla f(y)|^2 dy. \tag{1.1.4}$$

The next example shows that this last inequality fails to be true.

EXAMPLE 1.1.1:   Consider the function

$$f(x) = \begin{cases} \log|\log|x|| & \text{if } |x| \leq 1/e \\ 0 & \text{otherwise.} \end{cases}$$

Then $\|\nabla f\|_2^2 = 2\pi \int_0^{1/e} \frac{dr}{r|\log r|^2} = 2\pi$ but f is not bounded. Of course, f is not smooth, but it can easily be approximated by smooth functions f_n such that $\|\nabla f_n\|_2 \to \|\nabla f\|_2$ and $f_n \to f$. This shows that that (1.1.4) cannot be true.

What is true is recorded in the following theorem.

Theorem 1.1.1 *Fix an integer $n \geq 2$ and a real p, $1 \leq p < n$ and set $q = np/(n-p)$. Then there exists a constant $C = C(n, p)$ such that*

$$\forall\, f \in \mathcal{C}_0^\infty(\mathbb{R}^n), \quad \|f\|_q \leq C\|\nabla f\|_p. \tag{1.1.5}$$

This inequality is called the Sobolev inequality although the case $p = 1$ is not contained in [78]. Note that the case $p = n$ (i.e., $q = \infty$) is excluded in this result as should be the case according to the preceding example.

In the next few subsections we will give or outline several proofs of (1.1.5). As it turns out, when $p = 1$, (1.1.5) has a very simple proof based on

(1.1.1) and Hölder's inequality. This well-known proof (due independently to E. Gagliardo [28] and L. Nirenberg [68]) is presented in the next section. Moreover, as we shall see in 1.1.3 below, the case $p > 1$ follows from the case $p = 1$ by a simple trick.

We conclude this short introduction to Sobolev inequalities by recording a couple of useful remarks concerning the validity of (1.1.5). First, if (1.1.5) holds for all $f \in \mathcal{C}_0^\infty(\mathbb{R}^n)$, it obviously also holds for a larger class of functions including for instance all $\mathcal{C}^1$ functions with compact support or even Lipschitz functions vanishing at infinity. In fact, (1.1.5) holds for all functions vanishing at infinity whose gradient in the sense of distributions is in L^p. Second, (1.1.5) restricted to non-negative functions in $\mathcal{C}_0^\infty(\mathbb{R}^n)$ suffices to prove (1.1.5) in its full generality. Indeed, (1.1.5) for such functions implies that it also holds true for non-negative Lipschitz functions with compact support and, if $f \in \mathcal{C}_0^\infty(\mathbb{R}^n)$, $|f|$ is Lipschitz and satisfies $|\nabla|f|| \leq |\nabla f|$ almost everywhere. It then follows that (1.1.5) holds for $f \in \mathcal{C}_0^\infty(\mathbb{R}^n)$.

1.1.2   The proof due to Gagliardo and to Nirenberg

Recall that Hölder's inequality asserts that, for any positive measure μ,

$$\left| \int fg \, d\mu \right| \leq \|f\|_p \|g\|_{p'}$$

for all $f \in L^p(\mu)$, $g \in L^{p'}(\mu)$, $1 \leq p, p' \leq \infty$ with $1 = 1/p + 1/p'$. By a simple induction we find that

$$\left| \int f_1 f_2 \ldots f_k \, d\mu \right| \leq \|f_1\|_{p_1} \|f_2\|_{p_2} \ldots \|f_k\|_{p_k} \qquad (1.1.6)$$

for all $f_i \in L^{p_i}$, $1 \leq i \leq k$, $1 \leq p_i \leq \infty$, $1/p_1 + 1/p_2 + \cdots + 1/p_k = 1$.

Now, fix $f \in \mathcal{C}_0^\infty(\mathbb{R}^n)$. By (1.1.1), for any $x = (x_1, \ldots, x_n)$ and any integer $1 \leq i \leq n$, we have

$$|f(x)| \leq \frac{1}{2} \int_{-\infty}^{+\infty} |\partial_i f(x_1, \ldots, x_{i-1}, t, x_{i+1}, \ldots, x_n)| \, dt$$

(with the obvious interpretation if $i = 1$ or n). Set

$$F_i(x) = \int_{-\infty}^{+\infty} |\partial_i f(x_1, \ldots, x_{i-1}, t, x_{i+1}, \ldots, x_n)| \, dt$$

and

$$F_{i,m}(x) = \begin{cases} \int_{-\infty}^{+\infty} \cdots \int_{-\infty}^{+\infty} |\partial_i f(x)| \, dx_1 \ldots dx_m & \text{if } i \leq m \\ \int_{-\infty}^{+\infty} \cdots \int_{-\infty}^{+\infty} F_i(x) \, dx_1 \ldots dx_m & \text{if } i > m. \end{cases}$$

Note that each F_i depends only on $n-1$ variables, i.e., all coordinates but the ith. Similarly, $F_{i,m}$ depends on either $n-m$ or $n-m-1$ variables depending on whether $i \leq m$ or $i > m$. In particular, for $m = n$, $F_{i,n}(x) = \int_{\mathbb{R}^n} |\partial_i f(y)| dy$ is a constant function. Now, we can estimate f by

$$|f| \leq (1/2)(F_1 \ldots F_n)^{1/n}$$

so that

$$|f|^{n/(n-1)} \leq (1/2)^{n/(n-1)} (F_1 \ldots F_n)^{1/(n-1)}.$$

Using (1.1.6) with $k = n - 1$, $p_1 = p_2 = \cdots = p_k = n - 1$ and induction on $m \leq n$, one easily proves that

$$\int \cdots \int |f(x)|^{n/(n-1)} dx_1 \ldots dx_m \leq (1/2)^{n/(n-1)} (F_{1,m}(x) \ldots F_{n,m}(x))^{1/(n-1)}.$$

For $m = n$ this reads

$$\|f\|_{n/(n-1)} \leq (1/2) \left(\prod_1^n \|\partial_i f\|_1 \right)^{1/n}. \tag{1.1.7}$$

As $(\prod_1^n a_i)^{1/n} \leq \frac{1}{n} \sum_1^n a_i$ for any positive numbers a_i and integer n, we obtain

$$\|f\|_{n/(n-1)} \leq \frac{1}{2n} \sum_1^n \|\partial_i f(x)\|_1 dx \leq \frac{1}{2\sqrt{n}} \|\nabla f\|_1. \tag{1.1.8}$$

To see the last inequality, use $\sum_1^n |\partial_i f| \leq \sqrt{n} |\nabla f|$. This proves (1.1.5) for $p = 1$.

1.1.3 $p = 1$ implies $p \geq 1$

Assume that (1.1.5) holds for $p = 1$, that is,

$$\forall f \in \mathcal{C}_0^\infty(\mathbb{R}^n), \quad \|f\|_{n/(n-1)} \leq C\|\nabla f\|_1. \tag{1.1.9}$$

Fix $p > 1$. For any $\alpha > 1$ and $f \in \mathcal{C}_0^\infty(\mathbb{R}^n)$, note that $|f|^\alpha$ is $\mathcal{C}^1$, has compact support, and satisfies

$$|\nabla |f|^\alpha| = \alpha |f|^{\alpha-1} |\nabla f|.$$

Since we can easily approximate $|f|^\alpha$ by a sequence (f_i) of smooth functions with compact support such that $\nabla f_i \to \nabla |f|^\alpha$, inequality (1.1.9) holds with f replaced by $|f|^\alpha$. This yields

$$\begin{aligned}
\|f\|_{\alpha n/(n-1)}^\alpha &\leq C\alpha \int |f(x)|^{\alpha-1} |\nabla f(x)| dx \\
&\leq C\alpha \left(\int |f(x)|^{(\alpha-1)p'} dx \right)^{1/p'} \left(\int |\nabla f(x)|^p dx \right)^{1/p}
\end{aligned}$$

where $1/p + 1/p' = 1$. If we pick $\alpha = (n-1)p/(n-p)$, we find (rather miraculously) that $(\alpha - 1)q = n(p-1)p'/(n-p) = np/(n-p)$. Thus

$$\|f\|_{np/(n-p)}^{(n-1)p/(n-p)} \le C \frac{(n-1)p}{n-p} \|f\|_{np/(n-p)}^{n(p-1)/(n-p)} \|\nabla f\|_p.$$

Finally, $(n-1)p/(n-p) - n(p-1)/(n-p) = 1$, so that simplifying the last inequality yields

$$\|f\|_{np/(n-p)} \le C \frac{(n-1)p}{n-p} \|\nabla f\|_p.$$

Thus we have proved the following version of Theorem 1.1.1.

Theorem 1.1.2 *For any integer $n \ge 2$ and real p, $1 \le p < n$, set $q = np/(n-p)$. Then*

$$\forall f \in \mathcal{C}_0^\infty(\mathbb{R}^n), \quad \|f\|_{np/(n-p)} \le \frac{(n-1)p}{2(n-p)\sqrt{n}} \|\nabla f\|_p.$$

The Sobolev constant given by this theorem (i.e., the constant appearing in front of $\|\nabla f\|_p$) is not the best possible constant. This will be discussed in Section 1.3.1 below.

1.2 Riesz potentials

1.2.1 Another approach to Sobolev inequalities

Sobolev inequalities relate the size of ∇f to the size of f. In order to prove such inequalities, one may try to express f in terms of its gradient. We now derive such a representation formula. Using polar coordinates (r, θ), $r > 0$, $\theta \in \mathbb{S}^{n-1}$, in $\mathbb{R}^n$, write

$$f(x) = -\int_0^\infty \partial_r f(x + r\theta) dr$$

for any $f \in \mathcal{C}_0^\infty(\mathbb{R}^n)$. Integrating over the unit sphere $\mathbb{S}^{n-1}$ yields

$$\begin{aligned} f(x) &= -\frac{1}{\omega_{n-1}} \int_{\mathbb{S}^{n-1}} \int_0^\infty \partial_r f(x + r\theta) dr d\theta \\ &= -\frac{1}{\omega_{n-1}} \int_{\mathbb{S}^{n-1}} \int_0^\infty \frac{\partial_r f(x + r\theta)}{r^{n-1}} r^{n-1} dr d\theta. \end{aligned}$$

Here ω_{n-1} is the $(n-1)$-dimensional volume of the unit sphere $\mathbb{S}^{n-1} \subset \mathbb{R}^n$. That is, if Ω_n is the volume of the unit ball,

$$\omega_{n-1} = n\Omega_n = 2\pi^{n/2}/\Gamma(n/2)$$

where Γ is the gamma function ($\Gamma(n) = (n+1)!$ when n is an integer). Now, if $y = x + r\theta$, we have $r = |y - x|$ and

$$dy = r^{n-1}drd\theta \quad \text{and} \quad \partial_r f(x + r\theta) = |y - x|^{-1}\sum_1^n (y_i - x_i)\partial_i f(y).$$

Hence

$$f(x) = \frac{1}{\omega_{n-1}}\int_{\mathbb{R}^n} \frac{\langle x - y, \nabla f(y)\rangle}{|y - x|^n}dy. \tag{1.2.1}$$

In particular

$$|f(x)| \leq \frac{1}{\omega_{n-1}}\int_{\mathbb{R}^n} \frac{|\nabla f(y)|}{|y - x|^{n-1}}dy. \tag{1.2.2}$$

In view of this formula, we are led to study the properties of the convolution operator associated with $x \mapsto |x|^{-n+1}$.

More generally, for $0 < \alpha < n$, consider the Riesz potential operator I_α defined on $\mathcal{C}_0^\infty(\mathbb{R}^n)$ by

$$I_\alpha f(x) = \frac{1}{c_\alpha}\int_{\mathbb{R}^n} \frac{f(y)}{|y - x|^{n-\alpha}}dy \tag{1.2.3}$$

where $c_\alpha = \pi^{n/2}2^\alpha\Gamma(\alpha/2)/\Gamma((n - \alpha)/2)$. By Fourier transform arguments, one verifies that

$$I_\alpha f = \Delta^{-\alpha/2}f$$

where $\Delta = -\sum_1^n \partial_i^2 f$ is the Laplace operator. Here, $\Delta^{-\alpha/2}$ is defined using Fourier analysis. Namely, for all functions f in $\mathcal{C}_0^\infty(\mathbb{R}^n)$,

$$\widehat{\Delta^{\beta/2}f} = (2\pi|x|)^\beta \hat{f}, \quad \hat{f}(x) = \int_{\mathbb{R}^n} e^{2\pi i x \cdot y}f(y)dy.$$

The identity $I_\alpha f = \Delta^{-\alpha/2}f$ amounts to the fact that the Fourier transform of $c_\alpha^{-1}|x|^{-n+\alpha}$ is precisely $(2\pi|\xi|)^{-\alpha}$ in the sense that

$$c_\alpha^{-1}\int_{\mathbb{R}^n} |x|^{-n+\alpha}f(x)dx = \int_{\mathbb{R}^n} (2\pi|\xi|)^{-\alpha}\hat{f}(\xi)d\xi$$

for all $f \in \mathcal{C}_0^\infty(\mathbb{R}^n)$, when $0 < \alpha < n$. The restriction $0 < \alpha < n$ corresponds to the requirement that both $|x|^{-n+\alpha}$ and $|x|^{-\alpha}$ must be locally integrable for the above identity to make sense. One can show that $I_\alpha I_\beta = I_{\alpha+\beta}$ for $\alpha, \beta > 0$, $\alpha + \beta < n$, and $\Delta I_\alpha f = I_\alpha \Delta f = I_{\alpha-2}f$ for $2 \leq \alpha < n$.

Theorem 1.2.1 *Fix* $0 < \alpha < n$, $1 < p < n/\alpha$ *and define* q *by* $1/q = 1/p - \alpha/n$, *i.e.,* $q = np/(n - \alpha p)$. *Then there exists a constant* $C = C(n, \alpha, p)$ *such that*

$$\forall f \in \mathcal{C}_0^\infty(\mathbb{R}^n), \quad \|I_\alpha f\|_q \leq C\|f\|_p.$$

This theorem will be proved below in a more general form. As a corollary
of Theorem 1.2.1 and (1.2.2), we obtain (1.1.5) for $1 < p < n$. Observe
that the case $p = 1$ is excluded in Theorem 1.2.1. This has to be the case.
Indeed, if we had $\|I_\alpha f\|_q \le C\|f\|_1$, we could let $f \in \mathcal{C}_0^\infty(\mathbb{R}^n)$ tend to the
Dirac mass. This would imply that the function $x \mapsto |x|^{-n+\alpha}$ is in L^q with
$q = n/(n - \alpha)$. But this is clearly not the case.

Similarly, the case $p = n/\alpha$ must also be excluded. This follows from
the case $p = 1$ by duality, or more directly by the following example.

EXAMPLE   Consider

$$f(x) = \begin{cases} |x|^{-\alpha}(\log 1/|x|)^{-(\alpha/n)(1+\epsilon)} & \text{for } |x| \le 1/2 \\ 0 & \text{otherwise.} \end{cases}$$

Then $f \in L^{n/\alpha}$ if $\epsilon > 0$, but

$$I_\alpha f(0) = \frac{1}{c_\alpha} \int_{|x|<1/2} |x|^{-n}(\log|x|)^{-(\alpha/n)(1+\epsilon)} = \infty$$

when $\epsilon > 0$ is chosen small enough so that $(\alpha/n)(1 + \epsilon) < 1$ (recall that
$\alpha/n < 1$).

1.2.2   Marcinkiewicz interpolation theorem

Consider a measure space (M, μ). Fix $1 \le p, q \le \infty$. A linear operator K
defined on $L^1 \cap L^\infty$ is of weak type (p, q) if there exists a constant A such
that

$$\forall t > 0, \ \forall f \in L^1 \cap L^\infty, \ \mu(\{x : |Kf(x)| > t\}) \le (A\|f\|_p/t)^q. \qquad (1.2.4)$$

If $q = \infty$, this must be understood as $\|Kf\|_\infty \le A\|f\|_p$.

Theorem 1.2.2 *Assume that K is of weak types (p_1, q_1) and (p_2, q_2) with
$1 \le p_i, q_i \le \infty$, $p_1 < p_2$, $q_1 \ne q_2$. Then for each $0 < \theta < 1$ and $1/p =
\theta/p_1 + (1 - \theta)/p_2$, $1/q = \theta/q_1 + (1 - \theta)/q_2$, there exists a constant $C = C_\theta$
such that*

$$\|Kf\|_q \le C\|f\|_p.$$

This is an interpolation result due to Marcinkiewicz. It says that boundedness of K at the end points $(1/p_1, 1/q_1)$, $(1/p_2, 1/q_2)$ implies boundedness
along the segment joining these points in the $(1/p, 1/q)$ plane. See [79, 81]
for a proof. Of course, one of the important aspects here is that weak
boundedness at the end points is enough to prove strong boundedness in
the interior.

We want to apply this result to the case where K is given by a kernel $K(x, y)$ of weak type r for some $1 < r \leq \infty$, that is, which satisfies

$$\forall t > 0, \ \forall x, y \in M, \ \begin{cases} \mu(\{z : |K(x, z)| > t\}) \leq (A/t)^r, \\ \mu(\{z : |K(z, y)| > t\}) \leq (A/t)^r. \end{cases} \quad (1.2.5)$$

Again, if $r = \infty$, this must be understood as $\sup_{x,y} |K(x, y)| \leq A < \infty$.

Theorem 1.2.3 *Assume that*

$$Kf(x) = \int_M K(x, y) f(y) d\mu(y)$$

where K is a kernel of weak type r for some fixed $1 < r \leq \infty$. Then the operator K is of weak type (p, q) for all $1 \leq p < \infty$ and $p < q < \infty$ such that $1 + 1/q = 1/p + 1/r$. Moreover, for each such p, q, there exists a constant $B = B(r, p)$ such that

$$\forall f \in L^p, \ \|Kf\|_q \leq B\|f\|_p. \quad (1.2.6)$$

Without loss of generality we can assume that $K \geq 0$. For each $t > 0$, write $K = K_t + K^t$ where

$$K_t(x, y) = K(x, y) \mathbf{1}_{\{(u,v):K(u,v)\leq t\}}(x, y).$$

Lemma 1.2.4 *Fix $p \geq 1$. There exists a constant B_1 such that, for all $t > 0$ and all $f \in L^p$,*

$$\|K^t f\|_p \leq B_1 t^{-r+1} \|f\|_p.$$

Moreover, if $p/(p-1) < r$, there exists a constant B_2 such that, for all $t > 0$ and all $f \in L^p$,

$$\|K_t f\|_\infty \leq B_2 t^{1-r(p-1)/p} \|f\|_p.$$

To prove the first inequality, observe that

$$\int_M |K^t(x, y)| d\mu(y) = \int_0^\infty \mu(\{y : |K^t(x, y)| > s\}) ds$$

$$\leq \ t\mu(\{y : K(x, y) > t\}) + A \int_t^\infty s^{-r} ds \leq B_1 t^{1-r} \quad (1.2.7)$$

because $r > 1$. Thus $\|K^t f\|_\infty \leq B_1 t^{1-r}\|f\|_\infty$ and, by duality, $\|K^t f\|_1 \leq B_1 t^{1-r}\|f\|_1$. The duality argument runs as follows. For $f \in L^1 \cap L^\infty$, we have

$$\|K^t f\|_1 = \sup_{\substack{g \in L^\infty \\ \|g\|_\infty \leq 1}} \int (K^t f) \, g \, d\mu.$$

Moreover,

$$\int (K^t f)\, g\, d\mu = \int f\, (\widetilde{K}^t g)\, d\mu$$

where $\widetilde{K}(x,y) = K(y,x)$. From the x,y symmetry of our hypothesis it follows that (1.2.7) also holds for $\int_M |\widetilde{K}^t(x,y)| d\mu(y)$. Hence

$$\|\widetilde{K}^t g\|_\infty \le B_1 t^{1-r}$$

and

$$
\begin{aligned}
\int (K^t f)\, g\, d\mu &= \int f\, (\widetilde{K}^t g)\, d\mu \\
&\le \|\widetilde{K}^t g\|_\infty \|f\|_1 \\
&\le B_1 t^{1-r} \|g\|_\infty \|f\|_1.
\end{aligned}
$$

Thus

$$\|K^t f\|_1 = \sup_{\substack{g \in L^\infty \\ \|g\|_\infty \le 1}} \int (K^t f)\, g\, d\mu \le B_1 t^{1-r} \|f\|_1.$$

We still need to prove that

$$\|K^t f\|_p \le B_1 t^{1-r} \|f\|_p$$

for $1 < p < \infty$. To this end, use Jensen's inequality to obtain

$$|K^t f(x)|^p \le \left(\int_M K^t(x,y) d\mu(y) \right)^{p-1} \int_M K^t(x,y)|f(y)|^p d\mu(y).$$

Using the $L^1 \to L^1$ bound, we finally get $\|K^t f\|_p \le B_1 t^{1-r}\|f\|_p$ as desired.

To prove the second inequality of Lemma 1.2.4, write $1/p + 1/p' = 1$ so that $p' = p/(p-1)$ and note that

$$
\begin{aligned}
\int_M |K_t(x,y)|^{p'} d\mu(y) &= p' \int_0^\infty s^{p'-1} \mu(\{z : |K_t(x,z)| > s\}) ds \\
&\le p' A \int_0^t s^{p'-1-r} ds = p'(p'-r)^{-1} A\, t^{p'-r}
\end{aligned}
$$

because $p' < r$. It follows that

$$|K_t f| \le B_2 t^{1-r/p'} \|f\|_p.$$

To prove the first assertion of Theorem 1.2.3, fix $t > 0$ to be chosen later. Then, for any $s > 0$ and $f \in L^1 \cap L^\infty$ with $\|f\|_p = 1$, write

$$\mu(\{z : |Kf(z)| \ge s\}) \le \mu(\{z : |K_t f(z)| \ge s/2\}) + \mu(\{z : |K^t f(z)| \ge s/2\}).$$

By Lemma 1.2.4,

$$\mu(\{z : |K^t f(z)| \geq s/2\}) \leq (2\|K^t f\|_p/s)^p \leq (2B_1 t^{1-r}/s)^p$$

and

$$|K_t f| \leq B_2 t^{1-r(p-1)/p}.$$

Pick t so that $B_2 t^{1-r(p-1)/p} = s/4$. Thus $t = (B_2 s/4)^{p/(p+r-rp)}$. Then

$$\mu(\{z : |K^t f(z)| \geq s/2\}) = 0$$

and

$$
\begin{aligned}
\mu(\{z : |K f(z)| \geq s\}) &\leq \mu(\{z : |K_t f(z)| \geq s/2\}) \leq (2B_1 t^{1-r}/s)^p \\
&\leq B_3 \, s^{-p[1-(1-r)p/(p+r-rp)]} \\
&= B_3 \, s^{-pr/(p+r-rp)} = B_3 \, s^{-q}
\end{aligned}
$$

if $1/q = 1/p + 1/r - 1$, that is $q = pr/(p + r - rp)$. In words, the operator K is of weak type (p, q). This is true for all $1 < p < \infty$. The last assertion of Theorem 1.2.3 now follows from the Marcinkiewicz interpolation theorem, i.e., Theorem 1.2.2 with $1 < p_1 < p_2 < \infty$ arbitrary and $1/q_i = 1/p_i + 1/r - 1$. This ends the proof of Theorem 1.2.3. This is a typical use of the Marcinkiewicz interpolation theorem. We have turned a weak (p, q) boundedness result into a strong (p, q) boundedness result using the fact that the weak result holds for all p in a certain interval.

1.2.3   Proof of Sobolev Theorem 1.2.1

In order to prove Theorem 1.2.1, note that $K(x, y) = |x - y|^{-n+\alpha}$ is of weak type $n/(n - \alpha)$. Hence, by Theorem 1.2.3, I_α is of weak type $(1, n/(n - \alpha))$ and satisfies $\|I_\alpha f\|_q \leq C\|f\|_p$ for all $1 < p < \infty$ with $1/q = 1/p - \alpha/n$.

1.3   Best constants

1.3.1   The case $p = 1$: isoperimetry

Let $\mathbb{B}_n(r)$ and $\mathbb{S}^{n-1}(r)$ denote respectively the ball and the sphere of radius r centered at the origin in $\mathbb{R}^n$. Let $\Omega_n = \mu_n(\mathbb{B}^n(1))$ and $\omega_{n-1} = \mu_{n-1}(\mathbb{S}_{n-1}(1))$. The isoperimetric inequality in $\mathbb{R}^n$ asserts that among sets having a smooth boundary of given finite $(n-1)$-dimensional measure, the ball has the largest n-dimensional volume. Namely,

$$\mu_n(\Omega) \leq \mu_n(\mathbb{B}^n(r)) = \Omega_n \, r^n$$

where r is such that

$$\mu_{n-1}(\partial\Omega) = \mu_{n-1}(\mathbb{S}^{n-1}(r)) = \omega_{n-1} r^{n-1},$$

that is
$$r = (\mu_{n-1}(\partial\Omega)/\omega_{n-1})^{1/(n-1)}.$$

Hence, for any compact set $\Omega \subset \mathbb{R}^n$ with smooth boundary,

$$[\mu_n(\Omega)]^{(n-1)/n} \leq C_n \mu_{n-1}(\partial\Omega) \qquad (1.3.1)$$

where

$$C_n = \frac{\Omega_n^{1-1/n}}{\omega_{n-1}} = \frac{[\Gamma((n-1)/2)]^{1/n}}{\sqrt{\pi}\,n}.$$

Indeed, recall that $n\Omega_n = \omega_{n-1}$ and $\Omega_n = \pi^{n/2}/\Gamma((n-1)/2)$. This inequality has been known to geometers for a very long time; in particular, it was known well before Sobolev's work in the 1930's.

Apparently, the discovery that (1.3.1) is equivalent to Sobolev inequality (1.1.5) for $p = 1$ with the same constant, that is,

$$\forall f \in \mathcal{C}_0^\infty(\mathbb{R}^n), \ \|f\|_{n/(n-1)} \leq C_n\|\nabla f\|_1, \qquad (1.3.2)$$

was only made much later. In fact, in [78], Sobolev only proved (1.1.5) for $p > 1$. The case $p = 1$ is attributed to Gagliardo and to Nirenberg who published the proof given in Section 1.1.2 in 1958 and 1959 respectively. The connection between (1.3.1) and (1.3.2) was made in 1960 by Maz'ja and by Federer and Fleming. See [60].

The fact that (1.3.1) follows from (1.3.2) is rather straightforward. One approximates the function $\mathbf{1}_\Omega$ by smooth functions f_n so that

$$\|f_n\|_{n/(n-1)} \to \mu_n(\Omega)^{(n-1)/n} \text{ and } \|\nabla f_n\|_1 \to \mu_{n-1}(\partial\Omega).$$

To prove the other direction one needs the following co-area formula. See, e.g., [60, 1.2.4] and the references therein.

Theorem 1.3.1 *For any $f, g \in \mathcal{C}_0^\infty(\mathbb{R}^n)$,*

$$\int g|\nabla f|d\mu_n = \int_{-\infty}^{+\infty} \left(\int_{f(x)=t} g(x)d\mu_{n-1}(x) \right) dt.$$

Indeed, with this theorem at hand, for any smooth compactly supported $f \geq 0$ we have

$$\begin{aligned}
\int |f(x)|^{n/(n-1)}dx &\leq \int_0^\infty \mu_n(\{f > t\})^{(n-1)/n}dt \\
&\leq C_n \int_0^\infty \mu_{n-1}(\{f = t\})dt \\
&= C_n \int |\nabla f|d\mu_n = \|\nabla f\|_1.
\end{aligned}$$

To see the first inequality, write

$$f(x) = \int_0^\infty \mathbf{1}_{\{f(x)>t\}}(t)dt$$

and use the Minkowski inequality

$$\left\| \int f(\cdot,y)dy \right\|_q \leq \int \|f(\cdot,y)\|_q dy$$

with $q = n/(n-1) > 1$ to obtain

$$\left\| \int_0^\infty \mathbf{1}_{\{f(\cdot)>t\}}(t)dt \right\|_{n/(n-1)} \leq \int_0^\infty \|\mathbf{1}_{\{f(\cdot)>t\}}\|_{n/(n-1)}dt$$

$$= \int_0^\infty \mu_n(\{z : f(z) > t\})^{(n-1)/n}dt.$$

This shows that the isoperimetric inequality (1.3.1) implies the Sobolev inequality (1.3.2) with the same constant C_n.

1.3.2   A complete proof with best constant for $p = 1$

According to M. Gromov [62], inequality (1.3.2) goes back to H. Brunn's inaugural dissertation in 1887. My understanding is that Brunn proved the celebrated Brunn-Minkowski inequality (for convex sets and without the case of equality due to Minkowski) from which the isoperimetric inequality easily follows. Whether or not Brunn dicussed the isoperimetric inequality is not clear to me. Of course, he did not discuss (1.3.2). As mention earlier, the observation that (1.3.2) is equivalent to (1.3.1) is usually attributed to Maz'ja and to Federer and Fleming.

Gromov gives the following beautiful proof of (1.3.2) which he attributes to H. Knothe [48]. As we shall see, this proof yields both a proof of the isoperimetric inequality (1.3.1) and a proof of the (equivalent) Sobolev inequality (1.3.2). Again, as far as I can tell, there is no discussion of (1.3.2) in the work of Knothe.

Let g be a non-negative, locally integrable function with compact support S. For $x \in S$, set

$$A_i(x) = \{z : z_j = x_j \text{ for } j < i \text{ and } z_i \leq x_i\}$$

and

$$B_i(x) = \{z : z_j = x_j \text{ for } j < i\}$$

with the convention that $B_1 = \mathbb{R}^n$. Consider the map $y_g : x \mapsto y$ defined by

$$y_i = \int_{A_i(x)} g(z)dz_i \dots dz_n \left/ \int_{B_i(x)} g(z)dz_i \dots dz_n. \right.$$

Since $g \geq 0$ and $A_i \subset B_i$, we have $0 \leq y_i \leq 1$ for all x. That is, y is a map from S to the cube $[0,1]^n$. Obviously, this map is triangular, that is, for each i, y_i is a function of $x_1, \ldots, x_i$ only. Clearly, for each i, the partial derivative $\partial y_i / \partial x_i$ is non-negative and equal to

$$\frac{\partial y_i}{\partial x_i}(x) = \begin{cases} \int_{B_{i+1}(x)} g(z)dz_{i+1}\ldots dz_n \Big/ \int_{B_i(x)} g(z)dz_i \ldots dz_n & \text{if } 1 \leq i < n \\ g(x) \Big/ \int_{B_n(x)} g(z)dz_n & \text{if } i = n \end{cases}$$

at each point x where g is continuous. Thus, at any such point, the Jacobian of $x \mapsto y$ is equal to

$$J(x) = \prod_1^n \frac{\partial y_i}{\partial x_i} = g(x) \Big/ \int\int g(z)dz.$$

First, apply this construction to the case where $g = \chi$ is the characteristic function of the unit ball and observe that $y_\chi : \mathbb{B}^n \to (0,1)^n$ is invertible. Let $z = y_\chi^{-1}$ be the inverse map, $z : (0,1)^n \to \mathbb{B}^n$. Clearly, this is a triangular map with Jacobian equal to $\Omega_n = \mu_n(\mathbb{B}^n)$ in $[0,1]^n$.

The fact that y_χ is invertible is one of the crucial points of this proof. In fact, y_χ is invertible as soon as χ is the characteristic function of a convex set. We urge readers to check this for themselves.

Now, fix $f \in \mathcal{C}_0^\infty(\mathbb{R}^n)$, $f \geq 0$ and set $g = f^{n/(n-1)}$. We can assume that $\int g(x)dx = 1$. Construct the map y_g as above and consider the map

$$F = z \circ y_g : S = \text{supp}(f) \to \mathbb{B}^n.$$

By construction, the map F has non-negative partial derivatives $\partial F_i / \partial x_i$ and its Jacobian satisfies

$$J_F(x) = \Omega_n\, g(x)$$

for all x in the interior of S. Thus, the divergence $\text{div}F = \sum_1^n \partial F_i/\partial x_i$ satisfies

$$\frac{1}{n}\, \text{div}F(x) \geq [J_F(x)]^{1/n} = [\Omega_n\, g(x)]^{1/n}. \tag{1.3.3}$$

Furthermore, by the divergence theorem, we have

$$\int f(x)\text{div}F(x)dx = -\int \langle \nabla f(x), F(x)\rangle dx \leq \int |\nabla f(x)|dx \tag{1.3.4}$$

because $|F| \leq 1$. Hence

$$\begin{aligned} 1 = \int f(x)^{n/(n-1)}dx &= \int f(x)g(x)^{1/n}dx \\ &\leq \frac{1}{n\Omega_n^{1/n}} \int f(x)\, \text{div}F(x)dx \\ &\leq \frac{1}{n\Omega_n^{1/n}} \int |\nabla f(x)|dx. \end{aligned}$$

Removing the normalization $\int g(x)dx = 1$, we finally obtain

$$\|f\|_{n/(n-1)} \leq \frac{1}{n\Omega_n^{1/n}}\|\nabla f\|_1$$

which is exactly (1.3.2).

The same argument gives (1.3.1) if we take f to be the characteristic function of a bounded domain Ω with smooth boundary and if we replace (1.3.4) by

$$\int_\Omega \mathrm{div}F d\mu_n = \int_{\partial\Omega} \langle F, \mathbf{n}\rangle d\mu_{n-1}$$

where $\mathbf{n}$ is the exterior normal along $\partial\Omega$.

1.3.3   The case $p > 1$

The following theorem gives the best constant in the Sobolev inequality for $1 \leq p < \infty$.

Theorem 1.3.2 *For $1 \leq p < n$, the Sobolev inequality*

$$\forall f \in \mathcal{C}_0^\infty(\mathbb{R}^n), \quad \|f\|_q \leq C\|\nabla f\|_p$$

holds with C equal to

$$C(n,p) = \frac{p-1}{n-p}\left(\frac{n-p}{n(p-1)}\right)^{1/q}\left(\frac{\Gamma(n+1)}{\Gamma(n/p)\Gamma(n+1-n/p)\omega_{n-1}}\right)^{1/n} \quad (1.3.5)$$

for $1 < p < n$ and

$$C(n,1) = \frac{1}{n}\left(\frac{n}{\omega_{n-1}}\right)^{1/n}. \quad (1.3.6)$$

These are the best possible constants and the functions

$$x \mapsto \left(a + b|x - y|^{p/(p-1)}\right)^{1-n/p},$$

$a, b > 0$, $y \in \mathbb{R}^n$, are the extremal functions when $1 < p < n$.

We only briefly sketch the proof. See [5, 85] for details.

The proof is in two steps. First, one shows that it suffices to treat the case where f is a non-negative radial decreasing function. This follows from the following classic rearrangement inequality. For any function $f \in \mathcal{C}_0^\infty(\mathbb{R}^n)$, $f \geq 0$, let f^* be the radial decreasing function such that

$$\mu_n(\{z : f^*(z) > t\}) = \mu_n(\{z : f(z) > t\}).$$

That is, $f^*(x) = f_*(|x|)$ where

$$f_*(t) = \sup\{s : \mu_n(\{z : f(z) > s\}) > \Omega_n t^n\},$$

$\Omega_n = \omega_{n-1}/n$ being the volume of the unit ball.

Theorem 1.3.3 *For all $f \in C_0^\infty(\mathbb{R}^n)$ and all $1 \le p < \infty$, we have*

$$\|\nabla f^*\|_p \le \|\nabla f\|_p.$$

One proof of this theorem uses the isoperimetric inequality (1.3.1) and the co-area formula of Theorem 1.3.1. Talenti [85] gives a nice account. See also [5, Proposition 2.17].

Theorem 1.3.3 reduces the proof of Theorem 1.3.2 to the following 1-dimensional statement.

Lemma 1.3.4 *Fix $1 \le p < n$ and set $q = pn/(n-p)$. Let h be a decreasing function which is absolutely continuous on $[0, \infty)$ and equal to zero at infinity. Then*

$$\left(\int_0^\infty |h(t)|^q t^{n-1} dt \right)^{1/q} \le C'(n,p) \left(\int_0^\infty |h'(t)|^p t^{n-1} dt \right)^{1/p}$$

where

$$
\begin{aligned}
C'(n,p) &= \frac{p-1}{n-p} \left(\frac{n-p}{n(p-1)} \right)^{1/q} \left(\frac{\Gamma(n+1)}{\Gamma(n/p)\Gamma(n+1-n/p)} \right)^{1/n} \\
&= C(n,p)\omega_{n-1}^{1/n}.
\end{aligned}
$$

Moreover, for $1 < p < n$, equality is attained for the functions $t \mapsto (a + bt^{p/(p-1)})^{1-n/p}$.

See [5, 85] for a proof and earlier references.

1.4 Some other Sobolev inequalities

1.4.1 The case $p > n$

What can be said about the size of smooth functions with compact support in terms of $\|\nabla f\|_p$ when $p > n$? The following theorem gives a partial answer.

Theorem 1.4.1 *For $p > n$ there exists a constant $C = C(n,p)$ such that for any set Ω of finite volume we have*

$$\forall f \in C_0^\infty(\Omega), \quad \|f\|_\infty \le C\mu_n(\Omega)^{1/n-1/p}\|\nabla f\|_p. \tag{1.4.1}$$

Start with (1.2.2), that is,

$$|f(x)| \le \frac{1}{\omega_{n-1}} \int_{\mathbb{R}^n} \frac{|\nabla f(y)|}{|x-y|^{n-1}} dy.$$

Define p' by $1/p + 1/p' = 1$ and note that $(n-1)(p'-1) = (n-1)/(p-1) < 1$. Let also R be such that $\mu_n(\Omega) = \mu_n(\mathbb{B}(R))$, that is $R = (\mu_n(\Omega)/\Omega_n)^{1/n}$.

Then use the following computation. Write

$$
\begin{aligned}
\int_\Omega \frac{1}{|x-y|^{p'(n-1)}} dy &\leq \int_{\mathbb{B}(R)} \frac{1}{|x-y|^{p'(n-1)}} dy \\
&\leq \omega_{n-1} \int_0^R r^{(1-n)p'+n-1} dr \\
&= \omega_{n-1}(1 - (n-1)(p'-1))^{-1} R^{1-(n-1)(p'-1)} \\
&= \omega_{n-1}(1 - (n-1)(p'-1))^{-1} R^{(p-n)/(p-1)} \\
&= \frac{\omega_{n-1}\mu_n(\Omega)^{(p-n)/n(p-1)}}{\Omega_n^{(p-n)/n(p-1)}(1 - (n-1)(p'-1))} \qquad (1.4.2) \\
&= B\,\mu_n(\Omega)^{(p-n)/n(p-1)}.
\end{aligned}
$$

Now, by (1.2.2),

$$
\begin{aligned}
\|f\|_\infty &\leq \left(\frac{1}{\omega_{n-1}} \int_\Omega \frac{1}{|x-y|^{p'(n-1)}} dy \right)^{1/p'} \|\nabla f\|_p \\
&\leq C\mu_n(\Omega)^{1/n - 1/p}\|\nabla f\|_p.
\end{aligned}
$$

This proves Theorem 1.4.1.

One crucial difference between Theorem 1.4.1 and Sobolev inequality (1.1.5) for $1 \leq p < n$ is that the right-hand side of (1.4.1) depends on the set Ω on which the function f is supported. As the volume of Ω tends to infinity, the term $\mu_n(\Omega)^{1/n - 1/p}$ also tends to infinity since $n < p$. In fact, when $n \leq p$, there is no way to control the size of f purely in terms $\|\nabla f\|_p$. To see this, consider the function $f_r : x \mapsto (r - |x|)_+$. This function is supported in $\mathbb{B}(r)$ and

$$
\|\nabla f_r\|_p = \mu_n(\mathbb{B}(r))^{1/p} = \Omega_n^{1/p} r^{n/p}.
$$

Also,

$$
\|f_r\|_q \geq \mu_n(\mathbb{B}(r/2))(r/2) = \Omega_n^{1/q}(r/2)^{n/q+1}.
$$

For any fixed q, the ratio $\|f_r\|_q / \|\nabla f\|_p$ tends to infinity as r tends to infinity if $n \leq p$.

Theorem 1.4.1 can be complemented as follows.

Theorem 1.4.2 *For $p > n$, there exists a constant $C = C(n,p)$ such that any function $f \in \mathcal{C}^\infty(\mathbb{R}^n)$ with $\|\nabla f\|_p < \infty$ satisfies*

$$
\sup_{\substack{x,y\in\mathbb{R}^n \\ x\neq y}} \left\{ \frac{|f(x) - f(y)|}{|x-y|^\alpha} \right\} \leq C\|\nabla f\|_p
$$

with $\alpha = 1 - n/p$.

Note that this result does not require that f vanishes at infinity. For the proof we need a localized version of the representation formula (1.2.2).

Lemma 1.4.3 *Let B be a ball of radius $r > 0$. Then,*

$$\forall f \in \mathcal{C}^{\infty}(B), \quad \forall x \in B, \quad |f(x) - f_B| \leq \frac{2^n}{\omega_{n-1}} \int_B \frac{|\nabla f(y)|}{|x - y|^{n-1}} dy$$

where

$$f_B = \frac{1}{\mu_n(B)} \int_B f(z) dz.$$

For $x, y \in B$ write

$$f(x) - f(y) = -\int_0^{|x-y|} \partial_\rho f\left(x + \rho \frac{y - x}{|y - x|}\right) d\rho.$$

It follows that

$$|f(x) - f(y)| \leq \int_0^{\infty} F\left(x + \rho \frac{y - x}{|y - x|}\right) d\rho$$

where

$$F(z) = \begin{cases} |\nabla f(z)| & \text{if } x \in B \\ 0 & \text{otherwise.} \end{cases}$$

Integrating with respect to $y \in B$ yields

$$\begin{aligned}
|f(x) - f_B| &= \left| f(x) - \frac{1}{\mu_n(B)} \int_B f(y) dy \right| \\
&\leq \frac{1}{\mu_n(B)} \int_B |f(x) - f(y)| dy \\
&\leq \frac{1}{\Omega_n r^n} \int_B dy \left\{ \int_0^{\infty} F\left(x + \rho \frac{y - x}{|y - x|}\right) d\rho \right\} \\
&\leq \frac{1}{\Omega_n r^n} \int_{\{y : |x - y| \leq 2r\}} dy \left\{ \int_0^{\infty} F\left(x + \rho \frac{y - x}{|y - x|}\right) d\rho \right\} \\
&= \frac{1}{\Omega_n r^n} \int_0^{\infty} \int_{\mathbb{S}^{n-1}} \int_0^{2r} F(x + \rho\theta) s^{n-1} ds d\theta d\rho \\
&= \frac{2^n}{n\Omega_n} \int_0^{\infty} \int_{\mathbb{S}^{n-1}} F(x + r\theta) d\theta dr \\
&= \frac{2^n}{\omega_{n-1}} \int_B \frac{|\nabla f(y)|}{|y - x|^{n-1}} dy.
\end{aligned}$$

This proves Lemma 1.4.3.

To obtain Theorem 1.4.2, it now suffices to apply Lemma 1.4.3 and the argument of the proof of Theorem 1.4.1 to obtain

$$|f(x) - f_B| \leq C\mu_n(B)^{1/n-1/p} \left(\int_B |\nabla f|^p d\mu_n \right)^{1/p} \leq C\mu_n(B)^{1/n-1/p} \|\nabla f\|_p$$

for all $x \in B$ and all balls $B \subset \mathbb{R}^n$. Thus, for all x, y such that $|x - y| \leq r$, we get

$$|f(x) - f(y)| \leq 2C\Omega_n r^{1-n/p} \|\nabla f\|_p \leq 2C\Omega_n |x - y|^{1-n/p} \|\nabla f\|_p.$$

This proves Theorem 1.4.2.

1.4.2   The case $p = n$

To treat the case $p = n$, we first compute

$$\int_\Omega \frac{1}{|x - y|^{r(n-1)}} dy$$

when $r < n/(n - 1)$. Actually, this has already been done in (1.4.2) where we have shown that

$$\int_\Omega \frac{1}{|x - y|^{r(n-1)}} dy \leq \frac{\omega_{n-1}}{1 - (r-1)(n-1)} [\mu_n(\Omega)/\Omega_n]^{-(n+r-nr)/n}. \qquad (1.4.3)$$

For any $n < q < \infty$, set $1/n - 1/q = \delta$ and $1/r = 1 + 1/q - 1/n = 1 - \delta$. Now, write

$$
\begin{aligned}
|f(x)| &\leq \frac{1}{\omega_{n-1}} \int \frac{|\nabla f(y)|}{|x - y|^{n-1}} dy \\
&\leq \frac{1}{\omega_{n-1}} \int \frac{|\nabla f(y)|^{n/q}}{|x - y|^{r(n-1)/q}} \times |\nabla f(y)|^{n\delta} \times \frac{1}{|x - y|^{r(n-1)(1-1/n)}} dy.
\end{aligned}
$$

Notice that $1/q + \delta + (1 - 1/n) = 1$, and use the Hölder inequality (1.1.6) with $p_1 = q$, $p_2 = 1/\delta$, $p_3 = n/(n - 1)$ to get

$$
\begin{aligned}
|f(x)| &\leq \frac{1}{\omega_{n-1}} \left(\int \frac{|\nabla f(y)|^n}{|x - y|^{r(n-1)}} dy \right)^{1/q} \times \\
&\quad \left(\int |\nabla f(y)|^n dy \right)^\delta \left(\int_{\text{supp}(f)} \frac{1}{|x - y|^{r(n-1)}} dy \right)^{1-1/n}.
\end{aligned}
$$

It follows that, if f is supported in Ω,

$$
\begin{aligned}
\|f\|_q &\leq \frac{1}{\omega_{n-1}} \|\nabla f\|_n^{n/q+n\delta} \left(\int_\Omega \frac{1}{|x - y|^{r(n-1)}} dy \right)^{1/q+1-1/n} \\
&\leq \frac{1}{\omega_{n-1}} \|\nabla f\|_n \left(\int_\Omega \frac{1}{|x - y|^{r(n-1)}} dy \right)^{1/r}.
\end{aligned}
$$

Thanks to (1.4.3), this yields

$$\|f\|_q \leq \frac{\omega_{n-1}^{-1+1/r}}{[1-(r-1)(n-1)]^{1/r}\Omega_n^{(n+r-nr)/nr}} \mu_n(\Omega)^{(n+r-nr)/nr} \|\nabla f\|_n.$$

As $1/r = 1 + 1/q - 1/n$, we get

$$\begin{aligned} \|f\|_q &\leq \frac{\omega_{n-1}^{1/q-1/n}}{[1-(r-1)(n-1)]^{1/r}\Omega_n^{1/q}} \mu_n(\Omega)^{1/q} \|\nabla f\|_n \\ &= \frac{n^{1/q}}{[1-(r-1)(n-1)]^{1/r}\omega_{n-1}^{1/n}} \mu_n(\Omega)^{1/q} \|\nabla f\|_n. \end{aligned}$$

Note that $1 - (r-1)(n-1) = n(n+1)/(nq+n-q) \geq (n+1)/q$ because $q > n$. Hence, for $q > n$, we get

$$\|f\|_q^q \leq q^{1+q(n-1)/n} \omega_{n-1}^{q/n} \mu_n(\Omega) \|\nabla f\|_n^q. \tag{1.4.4}$$

It follows that for any integer $k = n, n+1, \ldots,$

$$\int_\Omega \left(\frac{|f(x)|}{\|\nabla f\|_n}\right)^{kn/(n-1)} dx \leq [kn/(n-1)]^{1+k} \omega_{n-1}^{-k/(n-1)} \mu_n(\Omega).$$

For $k = 0, 1, \ldots, n-1$, the left-hand side is easily bounded by $C\mu_n(\Omega)$ by Jensen's inequality.

Clearly, the series

$$\sum_0^\infty \frac{\alpha^k k^k}{(k-1)!} \left(\frac{n}{(n-1)\omega_{n-1}^{1/(n-1)}}\right)^k$$

converges if $\alpha > 0$ is small enough, e.g., $\alpha < (n-1)\omega_{n-1}^{1/(n-1)}/en$, and for such α we have

$$\begin{aligned} \int_\Omega \exp\left(\alpha\left(\frac{|f(x)|}{\|\nabla f\|_n}\right)^{n/(n-1)}\right) dx &\leq \sum_0^\infty \frac{\alpha^k}{k!} \int_\Omega \left(\frac{|f(x)|}{\|\nabla f\|_n}\right)^{kn/(n-1)} dx \\ &\leq C\mu_n(\Omega). \end{aligned}$$

This inequality is often attributed to N. Trudinger [86] but it appears in an earlier paper of V. Yudovich [90]. See [61]. We record it in the following theorem.

Theorem 1.4.4 *There exist two constants $C_n, c_n > 0$ such that, for any $0 < c \leq c_n$, for any bounded set $\Omega \subset \mathbb{R}^n$ and any function $f \in \mathcal{C}_0^\infty(\Omega)$,*

$$\int_\Omega \exp\left[c\left(\frac{|f(x)|}{\|\nabla f\|_n}\right)^{n/(n-1)}\right] dx \leq C_n\mu_n(\Omega).$$

A more precise result is known [66]. Namely, let $\alpha_n = n\omega_{n-1}^{1/(n-1)}$. Then, for each bounded domain $\Omega \subset \mathbb{R}^n$,

$$\forall f \in \mathcal{C}_0^\infty(\Omega), \quad \int_\Omega \exp\left(\alpha\left(\frac{|f(x)|}{\|\nabla f\|_n}\right)^{n/(n-1)}\right) dx \leq C\mu_n(\Omega)$$

for all $0 < \alpha \leq \alpha_n$ whereas, if $\alpha > \alpha_n$,

$$\sup\left\{\int_\Omega \exp\left(\alpha|f(x)|^{n/(n-1)}\right) dx : f \in \mathcal{C}_0^\infty(\Omega), \ \|\nabla f\|_n = 1\right\} = \infty.$$

This is proved by reducing the problem to a 1-dimensional inequality (thanks to Theorem 1.3.3) and then studying this 1-dimensional inequality. See, e.g., [5].

1.4.3   Higher derivatives

This short section describes Sobolev inequalities involving higher order derivatives. Most of the results easily follow from the first order case, but some additional arguments are needed to obtain optimal statements.

For a function $f \in \mathcal{C}_0^\infty(\mathbb{R}^n)$, let

$$\nabla_k f = \left(\partial_{i_1} \ldots \partial_{i_k} f\right)_{(i_1,\ldots,i_k)},$$

$$|\nabla_k f| = \sqrt{\sum_{(i_1,\ldots,i_k)} |\partial_{i_1} \ldots \partial_{i_k} f|^2},$$

and

$$\|\nabla_k f\|_p = \left(\int_{\mathbb{R}^n} |\nabla_k f(x)|^p dx\right)^{1/p}$$

with the obvious interpretation if $p = \infty$. By induction based on the case $k = 1$ (which is treated in the previous sections), one easily obtains the following statement.

Theorem 1.4.5 *Fix two integers n, k, and $1 \leq p < \infty$.*

- *If $1 \leq kp \leq n$ and $q = np/(n - kp)$, there exists a constant $C = C(n, k, p)$ such that*

$$\forall f \in \mathcal{C}_0^\infty(\mathbb{R}^n), \quad \|f\|_q \leq C\|\nabla_k f\|_p.$$

- *If $kp = n$, there exist $c = c(n, k)$ and $C' = C(n, k)$ such that for all bounded subsets $\Omega \subset \mathbb{R}^n$,*

$$\forall f \in \mathcal{C}_0^\infty(\Omega), \quad \int_\Omega \exp\left[c\left(\frac{|f(x)|}{\|\nabla_k f\|_p}\right)^{n/(n-1)}\right] dx \leq C'\mu_n(\Omega).$$

- If $kp > n$, let m be the integer such that $m \leq k - n/p < m + 1$ and set $\alpha = k - n/p - m$. If $\alpha > 0$, there exists $B = B(n, k, p)$ such that for all $f \in C_0^\infty(\mathbb{R}^n)$ and all m-tuples $(i_1, \ldots, i_m)$,

$$\sup_{\substack{x, y \in \mathbb{R}^n \\ x \neq y}} \left\{ \frac{|\partial_{i_1} \ldots \partial_{i_m} f(x) - \partial_{i_1} \ldots \partial_{i_m} f(y)|}{|x - y|^\alpha} \right\} \leq B \|\nabla_k f\|_p.$$

The result given above for $kp = n$, $k \geq 2$ is not optimal. The optimal result is as follows.

Theorem 1.4.6 *If $kp = n$, there exist $c = c(n, k)$ and $C' = C(n, k)$ such that for all bounded subsets $\Omega \subset \mathbb{R}^n$,*

$$\forall f \in C_0^\infty(\Omega), \quad \int_\Omega \exp\left[c \left(\frac{|f(x)|}{\|\nabla_k f\|_p} \right)^{n/(n-k)} \right] dx \leq C' \mu_n(\Omega).$$

For the proof, recall the representation formula

$$f(x) = \int \langle P_{n,k}(x - y), \nabla_k f(y) \rangle d(y)$$

where $P_{n,k}$ is homogeneous of degree $-n + k$, $0 < k < n$. More precisely, using multi-indices notation, we have

$$f(x) = \frac{(-1)^k k}{\omega_{n-1}} \sum_{|\alpha|=k} \int_{\mathbb{R}^n} \frac{\theta^\alpha \partial_\alpha f(y)}{\alpha! |x - y|^{n-k}} dy$$

where $\theta = (\theta_i)_1^n = (x - y)/|x - y|$. Starting from this higher order representation formula and proceeding as for Theorem 1.4.4, one proves Theorem 1.4.6.

The case $p = 1$, $k = n$ is a very special case. Indeed, we obviously have

$$|f(x)| \leq \int_{-\infty}^{+\infty} \cdots \int_{-\infty}^{+\infty} |\partial_1 \ldots \partial_n f(y)| dy.$$

Hence

$$\|f\|_\infty \leq \|\nabla_n f\|_1.$$

This is (at last) the higher-dimensional version of inequality (1.1.1).

The statement given above in the case $kp > n$ is also not satisfactory because the case where n/p is an integer (i.e., $\alpha = 0$) is excluded. For instance the case $p = 2$, $k = 2$, $n = 2$ is not treated. When $n/p = \ell$ is an integer, the optimal result is as follows.

Theorem 1.4.7 *Fix $n < p < \infty$ with $n/p = \ell$ an integer. There exists a constant $C = C(n, k, p)$ such that, for any $f \in \mathcal{C}_0^\infty(\mathbb{R}^n)$ and any $(\ell-1)$-tuple $(i_1, \ldots, i_{\ell-1})$, the function $g = \partial_{i_1} \ldots \partial_{i_{\ell-1}} f$ satisfies*

$$\sup_{\substack{x,y\in\mathbb{R}^n \\ x\neq y}} \left\{ \frac{|g(x+y) + g(x-y) - 2g(x)|}{|y|} \right\} \leq C\|\nabla_k f\|_p. \tag{1.4.5}$$

This is proved by the technique discussed above with the help of the following inequality.

Lemma 1.4.8 *Fix $n > 2$. For any smooth function in a ball B, there exists a linear function P_f such that, for all $x \in B$,*

$$|f(x) - P_f(x)| \leq C(n) \int_B \frac{|\nabla_2 f(y)|}{|x-y|^{n-2}} dy. \tag{1.4.6}$$

To prove this, for any $x, y \in B$, write $y - x = \rho\theta$ and

$$
\begin{aligned}
f(y) - f(x) &= \int_0^\rho \partial_s f(x+s\theta) ds \\
&= -\int_0^\rho \partial_s^2 f(x+s\theta) s ds + \rho \partial_s f(x+s\theta)|_{s=\rho}.
\end{aligned}
$$

As $\partial_s f(x + s\theta) = \langle \theta, \nabla f(x + s\theta)\rangle$, this yields

$$|f(y) - f(x) - \langle y - x, \nabla f(y)\rangle| \leq \int_0^\rho |\nabla_2 f(x+s\theta)|\, s\, ds.$$

Setting

$$F(z) = \begin{cases} |\nabla_2 f(z)| & \text{if } z \in B \\ 0 & \text{otherwise} \end{cases}$$

and integrating in polar coordinates around x in the ball B gives

$$
\begin{aligned}
|f(x) - P_f(x)| &\leq \frac{1}{\Omega_n r^n} \int_0^{2r} \int_{\mathbb{S}^n} \int_0^\infty F(x+r\theta) s ds \rho^{n-1} d\rho \\
&\leq \frac{2^n}{\Omega_n} \int_B \frac{|\nabla_2 f(y)|}{|x-y|^{n-2}} dy
\end{aligned}
$$

where

$$P_f(x) = \frac{1}{\Omega_n r^n} \int_B [f(y) - \langle y - x, \nabla f(y)\rangle] dy.$$

With (1.4.6) at hand, the argument used for Theorem 1.4.2 yields

$$\forall x \in B, \quad \forall f \in \mathcal{C}^\infty(B), \quad |f(x) - P_f(x)| \leq C(n,p) r^{2-n/p} \|\nabla_2 f\|_p$$

where B is a ball of radius r and $2p > n$. Note that

$$P_f(x+y) + P_f(x-y) - 2P_f(x) = 0$$

because P_f is linear. Hence, for any $x, y \in B$,

$$|f(x+y) + f(x-y) - 2f(x)| \leq 4C(n,p)r^{2-n/p}\|\nabla_2 f\|_p.$$

For any $f \in \mathcal{C}^\infty(\mathbb{R}^n)$ and any x, y such that $|x - y| < r$, we can use this estimate in the ball of center x and radius r to obtain

$$|f(x+y) + f(x-y) - 2f(x)| \leq 4C(n,p)|x-y|^{2-n/p}\|\nabla_2 f\|_p.$$

This obviously gives (1.4.5) in the case where $k = 2$ and n/p is an integer with $0 < n/p < 2$, i.e., $n/p = 1$. The case where $k \geq 3$ follows from this by induction, using the first statement in Theorem 1.4.5.

1.5   Sobolev–Poincaré inequalities on balls

1.5.1   The Neumann and Dirichlet eigenvalues

Let Ω be an open bounded domain in $\mathbb{R}^n$ with smooth boundary $\partial\Omega$. Classically, one considers the following two eigenvalue problems:

(1) The Neumann eigenvalue problem

$$\begin{cases} \Delta u = \lambda u & \text{on } \Omega \\ \frac{\partial u}{\partial \mathbf{n}} = 0 & \text{on } \partial\Omega, \end{cases}$$

where $\mathbf{n}$ is the exterior normal along $\partial\Omega$.

(2) The Dirichlet eigenvalue problem

$$\begin{cases} \Delta u = \lambda u & \text{on } \Omega \\ u = 0 & \text{on } \partial\Omega. \end{cases}$$

The boundary condition in (1) (resp. (2)) is known as the Neumann (resp. Dirichlet) boundary condition. Solutions of these problems are pairs (u, λ) with u a smooth function and λ a real. In both cases, integrating $\Delta u = \lambda u$ against u on Ω with the normalization $\int_\Omega u^2 d\mu = 1$ and integrating by parts, we obtain

$$\begin{aligned} \lambda &= \lambda \int_\Omega u^2 d\mu = \int_\Omega u \Delta u \, d\mu \\ &= \int_\Omega |\nabla u|^2 d\mu + \int_{\partial\Omega} \frac{\partial u}{\partial \mathbf{n}} u \, d\mu_{n-1} \\ &= \int_\Omega |\nabla u|^2 d\mu \geq 0. \end{aligned}$$

For the Neumann problem, $u \equiv 1$, $\lambda = 0$, is an obvious solution. In view of this, it is natural to set

$$\lambda^N(\Omega) = \inf \left\{ \frac{\int_\Omega |\nabla u|^2 d\mu}{\int_\Omega u^2 d\mu} : u \neq 0, \ \int_\Omega u d\mu = 0, \ u \in \mathcal{C}^\infty(\Omega) \right\}$$

and

$$\lambda^D(\Omega) = \inf \left\{ \frac{\int_\Omega |\nabla u|^2 d\mu}{\int_\Omega u^2 d\mu} : u \neq 0, \ u \in \mathcal{C}_0^\infty(\Omega) \right\}.$$

Indeed, one can show that $\lambda^N(\Omega)$ (resp. λ^D) is the smallest real λ such that (1) (resp. (2)) admits a non-constant solution. Observe that, by definition, the inequality $\lambda^N \geq c$ (resp. $\lambda^D \geq c$) is equivalent to saying that, for all $u \in \mathcal{C}^\infty(\Omega)$ (resp. $u \in \mathcal{C}_0^\infty(\Omega)$),

$$\int_\Omega u^2 d\mu \leq \frac{1}{c} \int_\Omega |\nabla u|^2 d\mu.$$

This type of inequality is known as a (L^2) Poincaré inequality.

1.5.2   Poincaré inequalities on Euclidean balls

There are two sets of Poincaré inequalities on Euclidean balls, corresponding when $p = 2$ to the Dirichlet and Neumann eigenvalue problems.

Theorem 1.5.1 *Let $B = B(z, r)$ be a Euclidean ball of radius r and center z in $\mathbb{R}^n$. For any $1 \leq p < \infty$, we have*

$$\forall f \in \mathcal{C}_0^\infty(B), \quad \left(\int_B |f|^p d\mu \right)^{1/p} \leq r \left(\int_B |\nabla f|^p d\mu \right)^{1/p} \tag{1.5.1}$$

and also

$$\forall f \in \mathcal{C}^\infty(B), \quad \left(\int_B |f - f_B|^p d\mu \right)^{1/p} \leq 2^n r \left(\int_B |\nabla f|^p d\mu \right)^{1/p} \tag{1.5.2}$$

where $f_B = \mu(B)^{-1} \int_B f d\mu$ is the mean of f over B.

Clearly, we can assume that $B = \mathbb{B}$ is the unit ball. For the proof of (1.5.1), we use (1.2.2), that is

$$|f(x)| \leq \frac{1}{\omega_{n-1}} \int \frac{|\nabla f(y)|}{|y - x|^{n-1}} dy.$$

This yields

$$\int_{\mathbb{B}} |f| d\mu \leq \frac{1}{\omega_{n-1}} \int_{\mathbb{B}} |\nabla f(y)| \left(\int_{\mathbb{B}} \frac{dx}{|x - y|^{n-1}} \right) dy.$$

As

$$\int_{\mathbb{B}} \frac{dx}{|x-y|^{n-1}} \le \int_{\mathbb{B}} \frac{dx}{|x|^{n-1}} = \omega_{n-1},$$

we get

$$\int_{\mathbb{B}} |f| d\mu \le \int_{\mathbb{B}} |\nabla f| d\mu,$$

which is the case $p = 1$ of (1.5.1). The case $p > 1$ can be obtained in a number of ways. We will use Jensen's inequality for the measure $c(x)^{-1}|x - y|^{-n+1} \mathbf{1}_{\mathbb{B}}(y) dy$ where $x \in \mathbb{B}$ is fixed and $c(x) = \int_{\mathbb{B}} |x - y|^{1-n} dy$. Observe that $c(x) \le \omega_{n-1}$. By Jensen's inequality,

$$|f(x)|^p \le \frac{c(x)^{p-1}}{\omega_{n-1}^p} \int_{\mathbb{B}} \frac{|\nabla f(y)|^p}{|y-x|^{n-1}} dy \le \frac{1}{\omega_{n-1}} \int_{\mathbb{B}} \frac{|\nabla f(y)|^p}{|y-x|^{n-1}} dy.$$

Integrating over $x \in \mathbb{B}$ as in the case $p = 1$ gives the desired result. The proof of (1.5.2) is similar but uses Lemma 1.4.3 instead of (1.2.2). Let us note that the constants in (1.5.1), (1.5.2) are not optimal.

1.5.3 Sobolev–Poincaré inequalities

For any open set Ω and $1 \le p \le \infty$, we set

$$\|f\|_{p,\Omega} = \left(\int_{\Omega} |f|^p d\mu \right)^{1/p}.$$

With this notation the inequalities of Theorem 1.5.1 read

$$\forall f \in \mathcal{C}_0^\infty(\mathbb{B}), \ \|f\|_{p,B} \le r \|\nabla f\|_{p,B},$$

$$\forall f \in \mathcal{C}^\infty(B), \ \|f - f_B\|_{p,B} \le 2^n r \|\nabla f\|_{p,B}.$$

Sobolev inequalities localized in a given ball can be obtained by using the representation formula (1.2.1) and Lemma 1.4.3. Indeed, the kernel

$$K(x,y) = \mathbf{1}_B(x) \mathbf{1}_B(y) \frac{1}{|x-y|^{n-1}}$$

is a kernel of weak type $(n - 1)$ as defined in Section 1.2.2. Thus, the proof of Theorem 1.2.1 given in Section 1.2.3 applies here and yields the following inequalities. Note that the case $s = p$ in Theorem 1.5.2 below reduces to Theorem 1.5.1 and only the case $s = q$ needs to be proved. Note also that it suffices to treat the case where B is the unit ball.

Theorem 1.5.2 *Fix $1 \le p < n$ and set $q = np/(n - p)$. There exists a constant $C = C(n,p)$ such that for any smooth function with compact support in a ball $B \subset \mathbb{R}^n$ of radius $r > 0$, i.e., $f \in \mathcal{C}_0^\infty(B)$, we have*

$$\|f\|_{s,B} \le C \, r^{1+n(1/s-1/p)} \, \|\nabla f\|_{p,B} \tag{1.5.3}$$

for all $1 \leq s \leq q$.

When f is a smooth function on the ball B which does not necessarily vanish on the boundary, i.e., $f \in \mathcal{C}^\infty(B)$, we have instead

$$\|f - f_B\|_{s,B} \leq C\, r^{1+n(1/s-1/p)}\, \|\nabla f\|_{p,B} \qquad (1.5.4)$$

for all $1 \leq s \leq q$. Here, f_B is the mean of f over the ball B.

It is natural to wonder whether the ball B can be replaced by some more general bounded domain. Let Ω be a bounded domain in $\mathbb{R}^n$. On the one hand, there is no difficulty with the case of functions with compact support in Ω because any $f \in \mathcal{C}_0^\infty(\Omega)$ can be extended to a function in $\mathcal{C}_0^\infty(\mathbb{R}^n)$ by setting $f = 0$ outside Ω. Thus, Jensen's inequality and the usual Sobolev inequality in $\mathbb{R}^n$, i.e., $\forall f \in \mathcal{C}_0^\infty(\mathbb{R}^n)$, $\|f\|_q \leq C\|\nabla f\|_p$ with $q = np/(n-p)$, yield

$$\forall f \in \mathcal{C}_0^\infty(\Omega), \ \ \|f\|_{s,\Omega} \leq C\,\mu(\Omega)^{1/s-1/q}\, \|\nabla f\|_{p,\Omega} \qquad (1.5.5)$$

for all $1 \leq p < n$, $q = np/(n-p)$ and $1 \leq s \leq q$. If Ω has diameter d, we can bound $\mu(\Omega)$ in this inequality by $\Omega_n d^n$.

On the other hand, consider the problem of whether or not the inequality

$$\forall f \in \mathcal{C}^\infty(\Omega), \ \ \|f - f_\Omega\|_{p,\Omega} \leq C(p,\Omega)\|\nabla f\|_{p,\Omega} \qquad (1.5.6)$$

holds true for some finite constant $C(p,\Omega)$. It turns out that the answer depends in a subtle way on the regularity of the boundary of Ω. Inequality (1.5.6) does hold on domains with smooth (or even Lipschitz) boundary but there are domains on which it does not hold. The same is true for Sobolev–Poincaré inequalities of type (1.5.4) with $s > p$. The book [61] gives an account of what is known and references to the literature.

Finally, note that it is easy to treat the case of bounded convex domains by adapting the argument given above in the case of Euclidean balls. All the results described above hold for bounded convex domains with the radius r of B replaced by the diameter d of the domain.

Chapter 2

Moser's elliptic Harnack inequality

2.1 Elliptic operators in divergence form

2.1.1 Divergence form

Second order differential operators with possibly non-constant coefficients can be written in a number of different ways. In particular, if

$$L = -\sum_{i,j} a_{i,j}(x)\partial_i\partial_j + \sum_i c_i(x)\partial_i + c(x)$$

we can also write, if the $a_{i,j}$ are smooth functions,

$$L = -\sum_{i,j} \partial_i(a_{i,j}(x)\partial_j) + \sum_i b_i(x)\partial_i + c(x) \qquad (2.1.1)$$

where

$$b_i(x) = c_i(x) + \sum_\ell \partial_\ell a_{\ell,i}.$$

Denote by $A(x) = (a_{i,j}(x))_{1 \le i,j \le n}$ the matrix with entries $a_{i,j}(x)$, set $b(x) = (b_1(x), \ldots, b_n(x))$ and recall that

$$\mathrm{div} X = \sum_1^n \partial_i X_i$$

for any smooth map $X : \mathbb{R}^n \to \mathbb{R}^n$. Then L can be written as

$$Lf = -\mathrm{div}\,(A\nabla f) + \langle b, \nabla f \rangle + c.$$

In this case, we say that L is in divergence form.

The distinction between operators in divergence form and others is much more important than one would think at first sight. Distinct sets of analytical tools are used and different results are obtained for the two types of equations. Already, at a naive level, what it means to have no lower order terms (i.e., no terms of order zero or one) is different: the operator $L = -\sum_{i,j} a_{i,j}(x)\partial_i\partial_j$ has a lower order term when written in divergence form

$$L = -\sum_{i,j} \partial_i a_{i,j}\partial_j + \sum_j \left[\sum_i \partial_i a_{i,j}(x)\right]\partial_j.$$

Another way to witness the differences between these two forms of second order differential operators is to consider what happens when the coefficients $a_{i,j}$ are not smooth but simply measurable and bounded. For simplicity, we consider only the case of operators with no lower order terms. Let $a_{i,j}(x)$ be measurable bounded coefficients. On the one hand, it is clear that the operator

$$-\sum_{i,j} a_{i,j}(x)\partial_i\partial_j$$

is well defined on, say, $\mathcal{C}^2$ functions. On the other hand, making sense of

$$-\sum_{i,j} \partial_i(a_{i,j}(x)\partial_j)$$

requires some work. Actually, it is hard in general to describe a set of functions on which the latter operator acts in a reasonable sense. The clue is to define

$$-\sum_{i,j} \partial_i(a_{i,j}(x)\partial_j)$$

in a weak sense by saying that $-\sum_{i,j} \partial_i(a_{i,j}\partial_j f) = g$ for functions f,g locally in L^2 with $|\nabla f|$ locally in L^2 *if and only if*

$$\int_{\mathbb{R}^n} \sum_{i,j} a_{i,j}(x)\partial_i f(x)\partial_j \phi(x)dx = \int_{\mathbb{R}^n} g(x)\phi(x)dx$$

for all *test* functions ϕ in a suitably chosen space, e.g., $\mathcal{C}_0^\infty(\mathbb{R}^n)$.

In what follows we will study second order differential operators in divergence form satisfying a uniform ellipticity condition.

2.1.2   Uniform ellipticity

The aim of this chapter is to study certain properties of (weak) solutions of the equation

$$Lu = 0 \tag{2.1.2}$$

on a Euclidean ball B, when

$$Lf = -\sum_{i,j} \partial_i(a_{i,j}\partial_j f) \tag{2.1.3}$$

is in divergence form and the real matrix-valued function $A = (a_{i,j})$ is uniformly elliptic. This essential hypothesis means that there exists $0 < \lambda \leq 1$ such that

$$\forall x \in \mathbb{R}^n, \ \forall \xi, \zeta \in \mathbb{R}^n, \ \left\{ \begin{array}{ll} \lambda|\xi|^2 & \leq \ \sum_{i,j} a_{i,j}(x)\xi_i\xi_j, \\ \lambda^{-1}|\xi||\zeta| & \geq \ |\sum_{i,j} a_{i,j}(x)\xi_i\zeta_j|. \end{array} \right. \tag{2.1.4}$$

If $(a_{i,j})$ is symmetric, this means that its eigenvalues belong to $[\lambda, \lambda^{-1}]$. Observe in particular that (2.1.4) implies $\sum_{i,j} a_{i,j}\xi_i\xi_j \geq 0$.

It is helpful to introduce the following Sobolev spaces. Let $W_0^{1,2}(B)$ be the closure of $\mathcal{C}_0^\infty(B)$ for the norm $\sqrt{\|f\|_2^2 + \|\nabla f\|_2^2}$. Let $W^{1,2}(B)$ be the closure of $\mathcal{C}^\infty(B)$ for the same norm. Thus $W_0^{1,2}(B) \subset W^{1,2}(B)$. One can show that $W^{1,2}(B)$ is also the space of all functions in $L^2(B)$ whose first order partial derivatives in the sense of distribution can be represented by $L^2(B)$ functions. By definition, a (weak) solution of (2.1.2) in the ball B is a function $u \in W^{1,2}(B)$ such that

$$\forall \phi \in W_0^{1,2}(B), \ \int_{\mathbb{R}^n} \sum_{i,j} a_{i,j}(x)\partial_i u(x)\partial_j \phi(x)dx = 0. \tag{2.1.5}$$

A (weak) subsolution is a function $u \in W^{1,2}(B)$ satisfying

$$\int \sum_{i,j} a_{i,j}(x)\partial_i u(x)\partial_j \phi(x)dx \leq 0 \tag{2.1.6}$$

for all $\phi \in W_0^{1,2}(B)$, $\phi \geq 0$. Thus, u is a subsolution if $Lu \leq 0$ in the weak sense (2.1.6). A function u is a supersolution if $-u$ is a subsolution.

We will prove the following theorem.

Theorem 2.1.1 *Let L be as in (2.1.3) with $A = (a_{i,j})$ satisfying (2.1.4) for some $\lambda > 0$. For any $\delta \in (0,1)$, there exists $C = C(n, \lambda, \delta) > 0$ such that any positive solution u of (2.1.5) in a ball B satisfies the Harnack inequality*

$$\sup_{\delta B}\{u\} \leq C \inf_{\delta B}\{u\}. \tag{2.1.7}$$

Moreover, for any $\delta \in (0,1)$, there exist $C' = C'(n, \lambda, \delta) > 0$ and $\alpha = \alpha(n, \lambda, \delta) > 0$ such that any solution u of (2.1.5) in a ball B satisfies the Hölder continuity estimate

$$\sup_{x,y \in \delta B} \left\{ \frac{|u(x) - u(y)|}{|x - y|^\alpha} \right\} \leq C' r^{-\alpha} \|u\|_{\infty, B} \tag{2.1.8}$$

where r is the radius of B.

Here and in the sequel δB is the ball with the same center as B and radius equal to δ times the radius of B. The sup's and inf's above must be understood as essential sup's and inf's.

One of the important aspects of this result is that the constants C, C', α do not depend on the function u or on the ball B.

The technique that will be used below to prove Theorem 2.1.1 consists in taking advantage of the weak form (2.1.5) of the equation $Lu = 0$ by adequate choices of the test function ϕ. One of the key ideas is to use test functions ϕ of the form $u^\beta \psi^2$ where u is the studied weak solution of the equation under consideration and $\psi \in \mathcal{C}_0^\infty(B)$ is a genuine test function. Here, it will be useful to consider positive and negative β's. This wonderfully simple idea is surprising: using the unknown u to define the test function ϕ is a rather bold idea. It requires one to check that $u^\beta \psi^2$ is in $W_0^{1,2}(B)$. For positive u bounded away from 0 and $\beta \le 1$, this does work fine. For $\beta > 1$, this is fine if u is a positive *locally bounded* function in $W^{1,2}(B)$. However, requiring u to be locally bounded is not a reasonable a priori hypothesis when dealing with solutions (resp. subsolutions) of operators L with non-smooth coefficients. When one wants, as we do, to deal with the general case of measurable coefficients this difficulty can be resolved in two different ways:

(1) One may regularize the coefficients and show that the solutions of (2.1.5) can be approximated by solutions of similar uniformly elliptic equations with smooth coefficients. These solutions are smooth by local elliptic theory. It then suffices to prove the desired bounds for solutions of equations with smooth coefficients. Implementing this idea does require some work.

(2) Instead, one may work directly with the desired equation and its weak solutions in $W^{1,2}(B)$. Then, for $\beta > 1$, one must work with approximate power functions of the form $t \mapsto t^\beta$ for $0 < t \le T$, and $t \mapsto T^{\beta-1}t$ for $t > T$, and afterwards pass to the limit as T tends to infinity. Here we will take this second route which is much more direct. Of course, a posteriori, weak solutions (resp. subsolutions) of uniformly elliptic equations are locally bounded functions so that this difficulty disappears.

The following lemma is a good simple example of a result which requires the boundedness of the weak subsolution u in order to be meaningful. Indeed, use of this lemma should be avoided if one wants to deal directly with non-smooth coefficients, and we will not use it the sequel.

Lemma 2.1.2 *If u is a subsolution of (2.1.5) in B and $\epsilon \le u \le c$ for some $0 < \epsilon \le c < \infty$, then u^α is also a subsolution for all $\alpha \ge 1$.*

We have, for any $\phi \in \mathcal{C}_0^\infty(B)$,

$$\sum_{i,j} a_{i,j} \partial_i u^\alpha \partial_j \phi = \alpha \sum_{i,j} a_{i,j} u^{\alpha-1} \partial_i u \partial_j \phi$$

$$= \alpha \sum_{i,j} a_{i,j} \partial_i u \partial_j (u^{\alpha-1} \phi)$$

$$-\alpha(\alpha-1)\left(\sum_{i,j}a_{i,j}\partial_i u\partial_j u\right)u^{\alpha-2}\phi$$

$$\leq \ \alpha\sum_{i,j}a_{i,j}\partial_i u\partial_j(u^{\alpha-1}\phi).$$

Moreover, $u^{\alpha-1}\phi$ is in $W_0^{1,2}(B)$. Indeed, $u^{\alpha-1}\phi \in L^2(B)$ and

$$\nabla(u^{\alpha-1}\phi)=(\alpha-1)u^{\alpha-2}\phi\nabla u+u^{\alpha-1}\nabla\phi \in L^2(B)$$

since $\epsilon \leq u \leq c$. Hence,

$$\int_B\sum_{i,j}a_{i,j}\partial_i u^\alpha\partial_j\phi d\mu \leq \alpha\int_B\sum_{i,j}a_{i,j}\partial_i u\partial_j(u^{\alpha-1}\phi)d\mu \leq 0.$$

2.1.3 A Sobolev-type inequality for Moser's iteration

In this chapter we will use Sobolev inequalities to study some properties of the positive solutions of the equation $Lu=0$ using a technique known as Moser's iteration. We will use the Sobolev inequality in the following form. Let $\mathbb{B}$ be the unit ball in $\mathbb{R}^n$. For $n>2$, Theorem 1.5.2, (1.5.3), yields

$$\forall f\in\mathcal{C}_0^\infty(\mathbb{B}),\ \ \|f\|_{q,\mathbb{B}}\leq C_n\|\nabla f\|_{2,\mathbb{B}}$$

where $q=2n/(n-2)$. We also have the Hölder inequality,

$$\|f\|_{p,\mathbb{B}}\leq\|f\|_{q,\mathbb{B}}^\gamma\|f\|_{2,\mathbb{B}}^{1-\gamma}$$

for any $1\leq p\leq 2n/(n-2)$ and $0\leq\gamma\leq 1$ with

$$\frac{1}{p}=\frac{\gamma}{q}+\frac{1-\gamma}{2}.$$

For $\gamma=n/(n+2)$ this gives $p=2(1+2/n)$. Hence

$$\int|f|^{2(1+2/n)}d\mu\leq\|f\|_q^2\|f\|_2^{4/n}.$$

Combining with the Sobolev inequality above, we get

$$\forall f\in\mathcal{C}_0^\infty(\mathbb{B}),\ \ \int_\mathbb{B}|f|^{2(1+2/n)}d\mu\leq C_n^2\left(\int_\mathbb{B}|\nabla f|^2 d\mu\right)\left(\int_\mathbb{B}|f|^2 d\mu\right)^{2/n}.$$
$$(2.1.9)$$

This inequality is still valid, of course, for all functions with compact support in $\mathbb{B}$ and square integrable partial derivatives, i.e., for $f\in W_0^{1,2}(\mathbb{B})$.

For the case $n=1,2$, the above inequality is still valid (with the same proof) if we replace n in the inequality by any given number $\nu>2$ (see

Theorem 1.5.2). In particular, in dimension 1 or 2 we can use $\nu = 3$ and get the inequality

$$\int_{\mathbb{B}} |f|^{2(1+2/3)} d\mu \leq C^2 \left(\int_{\mathbb{B}} |\nabla f|^2 d\mu \right) \left(\int_{\mathbb{B}} |f|^2 d\mu \right)^{2/3} \qquad (2.1.10)$$

for all $f \in \mathcal{C}_0^\infty(\mathbb{B})$. In what follows, we will implicitly use (2.1.10) instead of (2.1.9) when the dimension (i.e., n) is 1 or 2.

The proofs given in the next few sections would work as well if we used the usual Sobolev inequality instead of (2.1.9). However, it is somewhat interesting to note that one can work with (2.1.9) which, a priori, is a weaker inequality. Moreover using (2.1.9) is technically convenient when one treats parabolic (versus elliptic) equations.

2.2 Subsolutions and supersolutions

2.2.1 Subsolutions

Let L be given by (2.1.3). Assume that the uniform ellipticity condition (2.1.4) is satisfied.

Let B be a ball in $\mathbb{R}^n$ with volume $V = \mu(B)$. In every proof below, we assume without loss of generality that B is the unit ball. Indeed, if B is a ball of radius r and center z then $v(x) = u(z+rx)$ is defined on the unit ball and satisfies $L_r v = 0$ where $L_r v = -\sum_{i,j} \partial_i \tilde{a}_{i,j} v$ with $\tilde{a}_{i,j}(x) = a_{i,j}(z + rx)$. This operator L_r is uniformly elliptic if and only if L is, and with the same constant λ. Note that this would not work if we had to worry about the size of the derivatives of the coefficients.

Let u be a positive subsolution of the equation $Lu = 0$ in B. All the constants C_i appearing in this section are independent of u and B.

Lemma 2.2.1 *There exists a constant $C_1 = C_1(n, \lambda)$ such that, for any $0 < \rho' < \rho \leq 1$ and $p \geq 2$,*

$$\int_{\rho' B} u^{p\theta} d\mu \leq C_1 (\rho - \rho')^{-2} V^{1-\theta} \left(p^2 \int_{\rho B} u^p d\mu \right)^\theta$$

with $\theta = 1 + 2/n$ if $n > 2$ and $\theta = 1 + 2/3$ for $n = 1, 2$.

By replacing u by $u + \epsilon$, $\epsilon > 0$, we can assume that u is bounded away from zero. The desired inequality is then obtained by letting ϵ tend to zero at the end.

Let us recall that our hypothesis is that

$$\forall \phi \in W_0^{1,2}(B), \quad \int_{\mathbb{R}^n} \sum_{i,j} a_{i,j}(x) \partial_i u(x) \partial_j \phi(x) d(x) \leq 0. \qquad (2.2.1)$$

Let $G : (0, \infty) \to (0, \infty)$ be a piecewise $\mathcal{C}^1$ function such that $G(s) = as$ for large s. Assume also that G has a non-decreasing, non-negative derivative $G'(u)$. Hence, G is non-decreasing and $G(s) \leq sG'(s)$. Finally, define $H(u) \geq 0$ by $H'(u) = \sqrt{G'(u)}$, $H(0) = 0$. Observe that $H(u) \leq uH'(u)$ as well.

Let ψ be a non-negative smooth function with compact support in B. Set $\phi = \psi^2 G(u)$. Then $\phi \geq 0$ and $\phi \in W_0^{1,2}(B)$ (here we use the fact that $G(s) = as$ for large s). Thus we can use ϕ in (2.2.1). We have

$$\sum_{i,j} a_{i,j} \partial_i u \partial_j \phi = \psi^2 G'(u) \sum_{i,j} a_{i,j} \partial_i u \partial_j u + 2\psi G(u) \sum_{i,j} a_{i,j} \partial_i u \partial_j \psi.$$

Hence, by (2.2.1),

$$\int_B \psi^2 G'(u) \sum_{i,j} a_{i,j} \partial_i u \partial_j u \, d\mu < 2 \left| \int_B \psi G(u) \sum_{i,j} a_{i,j} \partial_i u \partial_j \psi \, d\mu \right|.$$

Uniform ellipticity and the inequality $G(u) \leq uG'(u)$ give

$$\int_B \psi^2 G'(u) |\nabla u|^2 d\mu \leq 2\lambda^{-2} \int_B \psi u G'(u) |\nabla u| |\nabla \psi| d\mu.$$

The Cauchy–Schwarz inequality yields

$$\int_B \psi^2 G'(u) |\nabla u|^2 d\mu$$
$$\leq 2\lambda^{-2} \left(\int_B \psi^2 G'(u) |\nabla u|^2 d\mu \right)^{1/2} \left(\int_B u^2 G'(u) |\nabla \psi|^2 d\mu \right)^{1/2}.$$

Thus

$$\int_B \psi^2 G'(u) |\nabla u|^2 d\mu \leq 4\lambda^{-4} \int_B u^2 G'(u) |\nabla \psi|^2 d\mu.$$

As $\nabla(\psi H(u)) = \psi H'(u) \nabla u + H(u) \nabla \psi$ and

$$|\nabla \psi H(u)|^2 \leq 2(\psi^2 |H'(u)|^2 |\nabla u|^2 + H(u)^2 |\nabla \psi|^2)$$
$$\leq 2(\psi^2 G'(u) |\nabla u|^2 + u^2 G'(u) |\nabla \psi|^2),$$

we obtain

$$\int_B |\nabla \psi H(u)|^2 d\mu \leq 2(1 + 4\lambda^{-4}) \int_B u^2 G'(u) |\nabla \psi|^2 d\mu. \qquad (2.2.2)$$

Now, $\psi H(u)$ has compact support in B and is in $W_0^{1,2}(B)$ (this uses the fact that $G(s) = as$ for large s). Thus we can apply the Sobolev inequality

(2.1.9) which gives

$$\int_B |\psi H(u)|^{2(1+2/n)} d\mu \tag{2.2.3}$$

$$\leq \; C_n^2 \left(\int_B |\nabla \psi H(u)|^2 d\mu \right) \left(\int_B |\psi H(u)|^2 d\mu \right)^{2/n}$$

$$\leq \; 2C_n^2(1+4\lambda^{-4}) \left(\int_B |\nabla\psi|^2 |u|^2 G'(u) d\mu \right) \left(\int_B |\psi|^2 u^2 G'(u) d\mu \right)^{2/n}$$

$$\leq \; 2C_n^2(1+4\lambda^{-4}) \|\nabla\psi\|_\infty^2 \|\psi\|_\infty^{4/n} \left(\int_{\text{supp}(\psi)} u^2 G'(u) d\mu \right)^{1+2/n}. \tag{2.2.4}$$

Given $0 < \rho' < \rho < 1$, we pick $\psi \in \mathcal{C}_0^\infty(B)$ so that $0 \leq \psi \leq 1$, $\psi = 1$ in $\rho'B$, $\psi = 0$ in ρB and $|\nabla\psi| \leq 2/(\rho - \rho')$. Then, (2.2.4) yields

$$\int_{\rho'B} |H(u)|^{2\theta} d\mu \leq 8C_n^2(1+4\lambda^{-4})(\rho - \rho')^{-2} \left(\int_{\rho B} u^2 G'(u) d\mu \right)^\theta \tag{2.2.5}$$

with $\theta = 1 + 2/n$. Fix $p \geq 1$ and some large N. Set

$$H_N(s) = \begin{cases} s^{p/2} & \text{if } s \leq N \\ N^{(p/2)-1}s & \text{if } s > N. \end{cases}$$

Computing, we find that this H_N corresponds to taking

$$\begin{aligned} G_N(s) &= \int_0^s H'(t)^2 dt \\ &= \frac{p^2}{4(p-1)} \begin{cases} s^{p-1} & \text{if } s \leq N \\ \frac{4(p-1)}{p^2} N^{p-2}(s-N) + N^{p-1} & \text{if } s > N. \end{cases} \end{aligned}$$

For any $p > 2$, these G_N's have all the required properties. Moreover, as N tends to infinity, $H_N(s) \to s^{p/2}$, $G_N'(s) \to (p/2)^2 s^{p-2}$. Hence, (2.2.5) yields

$$\int_{\rho'B} u^{p\theta} d\mu \leq 2C_n^2(1+4\lambda^{-4})(\rho - \rho')^{-2} \left(p^2 \int_{\rho B} u^p d\mu \right)^\theta.$$

This proves Lemma 2.2.1.

Our next step is to prove an L^p mean value inequality.

Theorem 2.2.2 *There exists $C_2 = C_2(n, \lambda)$ such that, for any $0 < \delta < 1$, any $p \geq 2$, and any positive subsolution u of $Lu = 0$ in a ball B of volume V,*

$$\sup_{\delta B}\{u^p\} \leq C_2(1-\delta)^{-n} \left(V^{-1} \int_B u^p d\mu \right).$$

We first prove a somewhat weaker version where the constant C_2 depends also on $p \geq 2$. Fix $p \geq 2$ and $0 < \delta < 1$. For each integer $i \geq 0$, set $p_i = p\theta^i$, $\rho_0 = 1$,

$$\rho_i = 1 - (1 - \delta) \sum_{j=1}^{j=i} 2^{-j}, \quad i \geq 1.$$

Then $\rho_{i+1} - \rho_i = (1 - \delta)2^{-i-1}$, $p_{i+1} = p_i\theta$, and, by Lemma 2.2.1,

$$\int_{\rho_{i+1}B} u^{p_{i+1}} d\mu \leq C(1 - \delta)^{-2} 2^{2(i+1)} \left(p_i^2 \int_{\rho_i B} u^{p_i} d\mu \right)^\theta$$

or

$$\left(\int_{\rho_{i+1}B} u^{p_{i+1}} d\mu \right)^{1/p_{i+1}} \leq [C(1-\delta)^{-2}]^{1/p_{i+1}} 2^{2(i+1)/p_{i+1}} p_i^{2/p_i} \left(\int_{\rho_i B} u^{p_i} d\mu \right)^{1/p_i}$$

for $i = 0, 1, \ldots$ with $C = 2C_n^2(1 + 4\lambda^{-4})$. This yields

$$\left(\int_{\rho_{i+1}B} u^{p_{i+1}} d\mu \right)^{1/p_{i+1}} \leq \left[C(n)C(p) \left[C(1-\delta)^{-2} \right]^{(\sum_1^{i+1} \theta^{-j})} \int_B u^p d\mu \right]^{1/p}$$

where

$$C(n) = 2^{2(\sum_1^\infty j\theta^{-j})}, \quad C(p) = e^{2\sum_0^\infty \theta^{-i} \log(p\theta_i)}.$$

Observe that $\rho_i \to \delta$,

$$\sum_1^\infty \theta^{-j} = \theta^{-1}(1 - \theta^{-1})^{-1} = n/2$$

and

$$\lim_{p \to \infty} \|f\|_{p,B} = \|f\|_\infty.$$

Hence

$$\sup_{\delta B} \{u\} \leq \left(C(n)C(p)C^n(1 - \delta)^n \right)^{1/p} \|u\|_{p,B}.$$

This proves Theorem 2.2.2 when B is the unit ball and with a constant C_2 which depends on p. In any case, it shows that all positive subsolutions in B are locally bounded. It follows that we can repeat the proof of Lemma 2.2.1 with $G(t) = t^{p-1}$. If one then uses direct computations instead of the inequality $G(s) \leq sG'(s)$, one obtains a sharper version of Lemma 2.2.1 that reads

$$\int_{\rho'B} u^{p\theta} d\mu \leq C_1(\rho - \rho')^{-2} V^{1-\theta} \left(\int_{\rho B} u^p d\mu \right)^\theta.$$

This is sharper because the factor p^2 in front of the last integral has been removed. Based on this inequality the argument above proves Theorem 2.2.2 with a constant C_2 which is independent of $p \geq 2$.

Theorem 2.2.2 can be interpreted as a form of L^p mean value inequality for subsolutions of $Lu = 0$. From this point of view, it is natural to try to extend the result to the case where the right-hand side is the L^p mean of u, $0 < p < 2$. In particular, it is natural to ask whether or not the L^1 inequality

$$\sup_{\delta B}\{u\} \leq C(1-\delta)^{-n}\left(V^{-1}\int_B u\, d\mu\right)$$

holds for positive subsolutions. That it holds is a special case of the following statement.

Theorem 2.2.3 *Fix $0 < p \leq 2$. There exists $C_3 = C_3(n, \lambda, p)$ such that, for any $0 < \delta < 1$ and any positive subsolution u of $Lu = 0$ in a ball B of volume V,*

$$\sup_{\delta B}\{u\} \leq C_3(1-\delta)^{n/p}\left(V^{-1}\int_B u^p d\mu\right)^{1/p}.$$

This can be obtained from Theorem 2.2.2 by the following similar but different iterative argument. As usual, we assume that B is the unit ball.

Fix $\sigma \in (1/2, 1)$ and set $\rho = \sigma + (1-\sigma)/4$. Then Theorem 2.2.2 yields

$$\sup_{\sigma B}\{u\} \leq C(1-\sigma)^{-n/2}\|u\|_{2,\rho B}$$

for some $C = C(n, \lambda)$. Now, as $\|u\|_{2,B} \leq \|u\|_{\infty,B}^{1-p/2}\|u\|_{p,B}^{p/2}$ for any ball B, we get

$$\|u\|_{\infty,\sigma B} \leq J(1-\sigma)^{-n/2}\|u\|_{\infty,\rho B}^{1-p/2}$$

where $J = C\|u\|_{p,B}^{p/2}$.

Fix $\delta \in (1/2, 1)$ and set $\sigma_0 = \delta$, $\sigma_{i+1} = \sigma_i + (1-\sigma_i)/4$. Then $1 - \sigma_i = (3/4)^i(1-\delta)$. Applying the above inequality for each i yields

$$\|u\|_{\infty,\sigma_i B} \leq (4/3)^{ni/2}J(1-\delta)^{-n/2}\|u\|_{\infty,\sigma_{i+1}B}^{1-p/2}.$$

Hence, for $i = 1, 2, \ldots$,

$$\|u\|_{\infty,\delta B} \leq (4/3)^{(n/2)\sum_0^{i-1} j(1-p/2)^j}[J(1-\delta)^{-n/2}]^{\sum_0^{i-1}(1-p/2)^j}\|u\|_{\infty,\sigma_i B}^{(1-p/2)^i}.$$

Letting i tend to infinity yields

$$\|u\|_{\infty,\delta B} \leq (4/3)^{2n/p^2}[(1-\delta)^{-n/2}J]^{2/p},$$

that is,

$$\|u\|_{\infty,\delta B} \leq (4/3)^{2n/p^2}C^{2/p}(1-\delta)^{-n/p}\|u\|_{p,B}$$

which is the desired inequality.

2.2.2   Supersolutions

We now proceed to derive some inequalities for supersolutions of $Lu = 0$ with L as in (2.1.3) satisfying the uniform ellipticity condition (2.1.4). The first result we derive below is similar to Theorem 2.2.3 except that it deals with negative powers u^{-p} of a positive supersolution u. The second result of this section deals with small positive powers u^p of supersolutions. These two results (Theorems 2.2.4 and 2.2.5 below) are proved in a form that describes precisely how the corresponding inequalities depend on the parameter p. This will be crucial later on in our proof of Theorem 2.1.1.

In this subsection, we let B be a fixed ball with volume V and we let u be a positive supersolution of $Lu = 0$. In the proofs, we will always assume as we may that B is the unit ball. The different constants C_i appearing below do not depend on u or B.

Theorem 2.2.4 *There exists a constant $C_4 = C_4(n, \lambda)$ such that, for any $\delta \in (0, 1)$ and any $p \in (0, \infty)$, we have*

$$\sup_{\delta B}\{u^{-p}\} \leq C_4(1 - \delta)^{-n}\frac{1}{V}\int_B u^{-p}d\mu.$$

Before embarking on the proof, let us observe that a slightly weaker form of this result can be obtained by applying Theorems 2.2.2, 2.2.3 to the positive, bounded subsolutions $(u+\epsilon)^{-1}$, $\epsilon > 0$. The improvement offered by Theorem 2.2.4 is that the constant C_4 above is independent of p whereas the constant obtained by applying Theorem 2.2.3 would be $C_3^p = C(4/3)^{2n/p}$ which tends to infinity as p tends to zero. Later on we will use the full strength of this improvement to obtain Harnack inequalities through the use of Lemma 2.2.6 below.

By replacing u by $u + \epsilon$, $\epsilon > 0$ and letting ϵ tend to 0 at the end of the argument, we can assume that u is bounded away from 0. For any $\phi \in W_0^{1,2}(B)$, we have

$$\int_B \sum_{i,j} a_{i,j}\partial_i u\partial_j\phi d\mu \geq 0. \tag{2.2.6}$$

We set $\phi = -\beta u^{\beta-1}\psi^2$ with $\beta < 0$ and $\psi \in W_0^{1,2}(B)$, $\psi \geq 0$. As ϕ is in $W_0^{1,2}(B)$ and

$$\partial_i\phi = -2\beta u^{\beta-1}\psi\partial_i\psi - \beta(\beta - 1)u^{\beta-2}\psi^2\partial_i u,$$

we have

$$-2\beta \int \psi u^{\beta-1}\sum_{i,j} a_{i,j}\partial_i u\partial_j\psi d\mu - \beta(\beta - 1)\int \psi^2 u^{\beta-2}\sum_{i,j} a_{i,j}\partial_i u\partial_j u d\mu \geq 0.$$

If we set $w = u^{\beta/2}$, we have $\partial_i w = (\beta/2)u^{(\beta/2)-1}\partial_i u$ and the inequality above reads

$$-4\int \psi w \sum_{i,j} a_{i,j}\partial_i w \partial_j \psi d\mu - \frac{4(|\beta|+1)}{|\beta|}\int \psi^2 \sum_{i,j} a_{i,j}\partial_i w \partial_j w d\mu \geq 0.$$

Hence

$$\int \psi^2 \sum_{i,j} a_{i,j}\partial_i w \partial_j w d\mu \leq \left| \int \psi w \sum_{i,j} a_{i,j}\partial_i w \partial_j \psi d\mu \right|.$$

By uniform ellipticity, this yields

$$\int \psi^2 |\nabla w|^2 d\mu \leq \lambda^{-2}\int \psi w |\nabla w||\nabla \psi| d\mu$$

and thus, by the Cauchy–Schwarz inequality,

$$\int \psi^2 |\nabla w|^2 d\mu \leq \lambda^{-4}\int |\nabla \psi|^2 w^2 d\mu.$$

Finally, we write this in terms of ψw using $\nabla(\psi w) = w\nabla \psi + \psi \nabla w$ and get

$$\int |\nabla \psi w|^2 d\mu \leq 2(1 + \lambda^{-4})\int |\nabla \psi|^2 w^2 d\mu. \tag{2.2.7}$$

This is exactly analogous to (2.2.2), and by the argument used after (2.2.2) we obtain

$$\int_{\rho'B} |w|^{2\theta} d\mu \leq 4C_n^2(1 + \lambda^{-4})(\rho - \rho')^{-2}\left(\int_{\rho B} |w|^2 d\mu\right)^{\theta}$$

where $0 < \rho' < \rho < 1$ and $\theta = 1 + 2/n$. Returning to our supersolution u, we have

$$\int_{\rho'B} u^{\beta\theta} d\mu \leq 4C_n^2(1 + \lambda^{-4})(\rho - \rho')^{-2}\left(\int_{\rho B} u^\beta d\mu\right)^{\theta}$$

for all $\beta < 0$. This is analogous to Lemma 2.2.1, and the iterative steps of the proof of Theorem 2.2.2 yield the inequality stated in Theorem 2.2.4.

It is not possible in general to control the sup norm of a positive supersolution in δB in terms of some L^p norm over B. Still, one has the following weaker result.

Theorem 2.2.5 *Fix $0 < p_0 < \theta = 1 + 2/n$ ($\theta = 1 + 2/3$ if $n = 1, 2$). There exists a constant $C_5 = C_5(n, \lambda, p_0)$ such that for all $\delta \in (0, 1)$ and for all $p \in (0, p_0/\theta)$,*

$$\left(\frac{1}{V}\int_{\delta B} u^{p_0} d\mu\right)^{1/p_0} \leq [C_5(1-\delta)^{-2n+2}]^{1/p - 1/p_0}\left(\frac{1}{V}\int_B u^p d\mu\right)^{1/p}.$$

As usual, we can assume that u is bounded away from 0. This time, we take $\phi = \beta u^{\beta-1}\psi^2$ in (2.2.6) with $0 < \beta \le p_0/\theta < 1$ and we set $\epsilon = 1 - (p_0/\theta) > 0$. We also set $w = u^{\beta/2}$. As in the beginning of the proof of Theorem 2.2.4, we obtain

$$2\beta \int \psi u^{\beta-1} \sum_{i,j} a_{i,j}\partial_i u \partial_j \psi d\mu + \beta(\beta-1) \int \psi^2 u^{\beta-2} \sum_{i,j} a_{i,j}\partial_i u \partial_j u d\mu \ge 0.$$

This yields

$$\frac{1-\beta}{\beta} \int \psi^2 \sum_{i,j} a_{i,j}\partial_i w \partial_j w d\mu \le \left| \int \psi w \sum_{i,j} a_{i,j}\partial_i w \partial_j \psi d\mu \right|.$$

The last inequality, the definition of ϵ, the upper bound on β and uniform ellipticity yield

$$\epsilon \int \psi^2 |\nabla w|^2 d\mu \le \lambda^{-2} \int \psi w |\nabla \psi||\nabla w| d\mu.$$

As usual, we use this to deduce that

$$\int |\nabla \psi w|^2 d\mu \le 2(1 + \lambda^{-4}\epsilon^{-2}) \int |\nabla \psi|^2 w^2 d\mu.$$

This is (again) exactly analogous to (2.2.2) and by the argument used after (2.2.2) we obtain

$$\int_{\rho'B} |w|^{2\theta} d\mu \le 4C_n^2 (1 + \lambda^{-4}\epsilon^{-2})(\rho - \rho')^{-2} \left(\int_{\rho B} |w|^2 d\mu \right)^\theta$$

where $0 < \rho' < \rho < 1$ and $\theta = 1 + 2/n$. Returning to our supersolution u, we have

$$\int_{\rho'B} u^{\beta\theta} d\mu \le 4C_n^2(1 + \lambda^{-4}\epsilon^{-2})(\rho - \rho')^{-2} \left(\int_{\rho B} u^\beta d\mu \right)^\theta \qquad (2.2.8)$$

for all $0 < \beta < p_0(1 + 2/n)^{-1}$. This is analogous to Lemma 2.2.1, but the iterative steps that will now be used to finish the proof of Theorem 2.2.5 are somewhat different from those used in the proof of Theorem 2.2.2.

Define $p_i = p_0\theta^{-i}$ where $\theta = 1 + 2/n$. We first prove Theorem 2.2.5 for these values of p. Thus, fix $i \ge 1$, and apply (2.2.8) with $\beta = p_i\theta^{j-1}$, $j = 1, 2, \ldots, i$, and with $\rho' = \sigma_{j-1}$, $\rho = \sigma_j$, where $\sigma_0 = 1$ and $\sigma_{\ell-1} - \sigma_\ell = 2^{-\ell}(1 - \delta)$. Observe that $p_i\theta^{j-1} \le p_0(1 + 2/n)^{-1}$ for $j = 1, \ldots, i$ as required for the validity of (2.2.8). Hence, for all $j = 1, \ldots, i$,

$$\int_{\sigma_i B} u^{p_i\theta^j} d\mu \le C(1 - \delta)^{-2} 2^{2j} \left(\int_{\sigma_{i-1}B} u^{p_i\theta^{j-1}} d\mu \right)^\theta$$

where $C = 4C_n^2(1 + \lambda^{-4}\epsilon^{-2})$. This yields

$$\int_{\sigma_i B} u^{p_0} d\mu \leq 2^{\sum_0^{i-1}(i-j)\theta^j} [C(1-\delta)^{-2}]^{\sum_0^{i-1}\theta^j} \left(\int_B u^{p_i} d\mu\right)^{\theta^i}.$$

Finally, observe that

$$\sum_0^{i-1}(i-j)\theta^j \leq (n/2)^2\theta^{i-1} \leq (n/2)^3(\theta^i - 1) = (n/2)^3(p_0/p_i - 1),$$

$$\sigma_i = 1 - \left(\sum_1^i 2^{-j}\right)(1-\delta) > \delta,$$

and

$$\sum_0^{i-1}\theta^j = \frac{\theta^i - 1}{\theta - 1} = (n/2)(p_0/p_i - 1).$$

This gives

$$\left(\int_{\delta B} u^{p_0} d\mu\right)^{1/p_0} \leq [2^{2(n/2)^3}C(1-\delta)^{-2}]^{n(1/p_i-1/p_0)/2}\left(\int_B u^{p_i} d\mu\right)^{1/p_i},$$

that is,

$$\left(\int_{\delta B} u^{p_0} d\mu\right)^{1/p_0} \leq [C'(1-\delta)^{-n}]^{1/p_i-1/p_0}\left(\int_B u^{p_i} d\mu\right)^{1/p_i}$$

with $C' = 2^{2(n/2)^4}C^{n/2}$.

To obtain the desired inequality for any $p \in (0, p_0/\theta)$, let $i \geq 2$ be the integer such that $p_i \leq p < p_{i-1}$. Then $1/p_i - 1/p_0 \leq (1+\theta)(1/p - 1/p_0)$. Thus, by Jensen's inequality,

$$\begin{aligned}
\left(\int_{\delta B} u^{p_0} d\mu\right)^{1/p_0} &\leq [C'(1-\delta)^{-n}]^{1/p_i-1/p_0}\left(\int_B u^{p_i} d\mu\right)^{1/p_i}\\
&\leq [C'(1-\delta)^{-n}]^{(1+\theta)(1/p-1/p_0)}\left(\int_B u^p d\mu\right)^{1/p}\\
&\leq [C''(1-\delta)^{-(1+\theta)n}]^{1/p-1/p_0}\left(\int_B u^p d\mu\right)^{1/p}
\end{aligned}$$

with $C'' = (C')^{1+\theta}$.

2.2.3 An abstract lemma

This section presents an elementary but subtle lemma due to Bombieri and Giusti [9] which simplifies considerably Moser's original proof of the Harnack inequality. It replaces the use of the well-known John–Nirenberg inequality (i.e, the exponential integrability of BMO functions). This simplification is even more significant when dealing with parabolic equations. See [65].

Consider a collection of measurable subsets U_σ, $0 < \sigma \leq 1$, of a fixed measure space endowed with a measure μ, such that $U_{\sigma'} \subset U_\sigma$ if $\sigma' \leq \sigma$. In our application, U_σ will be σB for some fixed ball $B \subset \mathbb{R}^n$.

Lemma 2.2.6 *Fix $0 < \delta < 1$. Let γ, C be positive constants and $0 < \alpha_0 \leq \infty$. Let f be a positive measurable function on $U_1 = U$ which satisfies*

$$\|f\|_{\alpha_0, U_{\sigma'}} \leq \left[C(\sigma - \sigma')^{-\gamma} \mu(U)^{-1} \right]^{1/\alpha - 1/\alpha_0} \|f\|_{\alpha, U_\sigma},$$

for all σ, σ', α such that $0 < \delta \leq \sigma' < \sigma \leq 1$ and $0 < \alpha \leq \min\{1, \alpha_0/2\}$. Assume further that f satisfies

$$\mu(\log f > \lambda) \leq C\mu(U)\lambda^{-1}$$

for all $\lambda > 0$. Then

$$\|f\|_{\alpha_0, U_\delta} \leq A\mu(U)^{1/\alpha_0}$$

where A depends only on δ, γ, C, and a lower bound on α_0.

For the proof, assume without loss of generality that $\mu(U) = 1$ and set

$$\psi = \psi(\sigma) = \log(\|f\|_{\alpha_0, U_\sigma}), \quad \text{for } 0 < \delta \leq \sigma \leq 1.$$

Decomposing U_σ into the sets where $\log f > \psi/2$ and where $\log f \leq \psi/2$, we get

$$
\begin{aligned}
\|f\|_{\alpha, U_\sigma} &\leq \|f\|_{\alpha_0, U_\sigma} \mu(\log f > \psi/2)^{1/\alpha - 1/\alpha_0} + e^{\psi/2} \\
&\leq e^\psi (2C/\psi)^{1/\alpha - 1/\alpha_0} + e^{\psi/2}.
\end{aligned}
\tag{2.2.9}
$$

Here, we have used successively the Hölder inequality and the second hypothesis of the lemma. We want to choose α so that the two terms in the right-hand side of (2.2.9) are equal, and $0 < \alpha \leq \min\{1, \alpha_0/2\}$. This is possible if

$$(1/\alpha - 1/\alpha_0)^{-1} = (2/\psi) \log(\psi/2C) \leq \min\{1, \alpha_0/2\},$$

and this last inequality is certainly satisfied when

$$\psi \geq A_1 \tag{2.2.10}$$

where A_1 depends only on a lower bound on α_0 (note that one can always take $C \geq 1$). Assuming that (2.2.10) holds and that α has been chosen as above, we obtain

$$\|f\|_{\alpha, U_\sigma} \leq 2e^{\psi/2}. \tag{2.2.11}$$

The first hypothesis of the lemma and (2.2.11) yield

$$\begin{aligned} \psi(\sigma') &\leq \log\left[\left(C(\sigma - \sigma')^{-\gamma}\right)^{1/\alpha - 1/\alpha_0} 2e^{\psi/2}\right] \\ &= (1/\alpha - 1/\alpha_0) \log\left(C(\sigma' - \sigma)^{-\gamma}\right) + \psi/2 + \log 2 \end{aligned}$$

for $\delta < \sigma' < \sigma \leq 1$. By our choice of α specified above we have

$$\psi(\sigma') \leq \frac{\psi}{2} \left[\frac{\log(C(\sigma - \sigma')^{-\gamma})}{\log(\psi/2C)} + 1 \right] + \log 2.$$

On the one hand, if

$$\psi \geq 2C^3(\sigma - \sigma')^{-2\gamma}, \tag{2.2.12}$$

we have

$$\psi(\sigma') \leq \frac{3}{4}\psi + \log 2.$$

On the other hand, if one of the hypotheses (2.2.10), (2.2.12) made on ψ is not satisfied, we have

$$\psi(\sigma') \leq \psi \leq A_1 + 2C^3(\sigma - \sigma')^{-2\gamma}.$$

In all cases, we obtain

$$\psi(\sigma') \leq \frac{3}{4}\psi(\sigma) + A_2(\sigma - \sigma')^{-2\gamma} \tag{2.2.13}$$

where A_2 depends only on C and on a lower bound on α_0.

For any sequence

$$0 < \delta = \sigma_0 < \sigma_1 < \cdots < \sigma_i \leq 1,$$

an iteration of (2.2.13) yields

$$\psi(\sigma_0) \leq (3/4)^i \psi(\sigma_i) + A_2 \sum_0^{i-1} (3/4)^j (\sigma_{j+1} - \sigma_j)^{-2\gamma},$$

and, when i tends to infinity,

$$\psi(\delta) \leq A_2 \sum_0^\infty (3/4)^j (\sigma_{j+1} - \sigma_j)^{-2\gamma}.$$

The desired bound follows if we set $\sigma_j = 1 - (1 + j)^{-1}(1 - \delta)$.

2.3 Harnack inequalities and continuity

2.3.1 Harnack inequalities

The main goal of this section is to prove the Harnack inequality of Theorem 2.1.1. In fact, we prove first a weaker inequality (Theorem 2.3.1 below) which holds for positive supersolutions. Then, the Harnack inequality of Theorem 2.1.1 immediately follows from this and Theorem 2.2.3.

Let L at (2.1.3) be a divergence form operator satisfying the uniform ellipticity condition (2.1.4).

Theorem 2.3.1 *Fix $\delta \in (0,1)$ and p, $0 < p < \theta = 1 + 2/n$ ($\theta = 1 + 2/3$ if $n = 1, 2$). There exists a constant $C = C(n, \lambda, \delta, p)$ such that for any ball B and any positive supersolution u of $Lu = 0$ in B we have*

$$\frac{1}{\mu(\delta B)} \int_{\delta B} u^p d\mu \le C \inf_{\delta B} \{u^p\}.$$

We assume that B is the unit ball. We want to apply Lemma 2.2.6 to $e^{-c}u$ and $e^c u^{-1}$ where c is a well-chosen constant. Pick ρ so that $\delta < \rho < 1$ (for instance $\rho = \delta + (1 - \delta)/2$). Set $U = B' = \rho B$, $V' = \mu(B')$, and set $U_\sigma = \sigma B' = \sigma \rho B$. Theorem 2.2.5 shows that the first hypothesis of Lemma 2.2.6 is satisfied by any constant multiple of u with $\alpha_0 = p < \theta$. To verify the second hypothesis of Lemma 2.2.6, we apply the L^2-Poincaré inequality in the ball $U = B'$ to the function $\log u$. This yields

$$\int_{B'} |\log u - c|^2 d\mu \le C_n \int_{B'} |\nabla \log u|^2 d\mu$$

where C_n depends only on n and

$$c = (\log u)_{B'} = \frac{1}{V'} \int_{B'} \log u \, d\mu$$

is the mean of $\log u$ over B'. Now, as u is a positive supersolution,

$$\int_B \sum_{i,j} a_{i,j} \partial_i u \partial_j \phi \ge 0$$

for all non-negative test functions $\phi \in W_0^{1,2}(B)$. Pick $\phi = u^{-1}\psi^2$ where ψ has support in B, $0 \le \psi \le 1$, $\psi = 1$ in B', $|\nabla \psi| \le 2/(1 - \rho)$. Then

$$\sum_{i,j} a_{i,j} \partial_i u \partial_j \phi = -\psi^2 \sum_{i,j} a_{i,j} \partial_i v \partial_j v + 2\psi \sum_{i,j} a_{i,j} \partial_i v \partial_j \psi$$

where $v = \log u$. Hence

$$\int \psi^2 \sum_{i,j} a_{i,j} \partial_i v \partial_j v d\mu \le 2 \left| \int \psi \sum_{i,j} a_{i,j} \partial_i v \partial_j \psi d\mu \right|.$$

Using uniform ellipticity and the Cauchy–Schwarz inequality as in the proof of Theorems 2.2.2, 2.2.4, 2.2.5, we get

$$\int \psi |\nabla v|^2 d\mu \le 2\lambda^{-4} \int |\nabla \psi|^2 d\mu \le 8\Omega_n (1-\rho)^{-2} \lambda^{-4} = C(n, \lambda, \delta).$$

Thus

$$
\begin{aligned}
t\mu(B' \cap \{|\log u - c| \ge t\}) &\le \int_{B'} |\log u - c| d\mu \\
&\le \left(\int_{B'} |\log u - c|^2 d\mu \right)^{1/2} \\
&\le C(n, \lambda, \delta).
\end{aligned}
$$

This shows that the second hypothesis of Lemma 2.2.6 is satisfied by $f = e^{-c}u$ (and also by $e^{c}u^{-1}$, for that matter). Hence, we can apply Lemma 2.2.6 to $f = e^{-c}u$ and conclude that, for any $0 < p < 1 + 2/n$,

$$\|u\|_{p,\delta B} \le Ae^c \tag{2.3.1}$$

with $A = A(n, \lambda, \delta)$. Likewise, Theorem 2.2.4 and the computation above show that Lemma 2.2.6 applies to $e^c u^{-1}$ with the *same* $c = (\log u)_{B'}$ as above. In this case, $\alpha_0 = \infty$. Hence

$$\sup_{\delta B}\{u^{-1}\} \le Ae^{-c}. \tag{2.3.2}$$

For $0 < p < 1 + 2/n$, putting (2.3.1) and (2.3.2) together yields

$$\|u\|_{p,\delta B} \le A^2 \inf_{\delta B}\{u\}$$

which is the desired inequality.

2.3.2   Hölder continuity

We now derive the last part of Theorem 2.1.1, that is, the Hölder continuity estimate

$$\sup_{x,y \in \delta B} \left\{ \frac{|u(x) - u(y)|}{|x - y|^\alpha} \right\} \le Cr^{-\alpha} \sup_B \{u\} \tag{2.3.3}$$

for positive solutions of $Lu = 0$ in the ball B of radius r. Let us mention that this result was first proved by De Giorgi in 1957, before the present proof was given by Moser in 1961. Also, Nash's paper [67], published in 1958, proves the Hölder continuity of the solutions of the related parabolic equation.

To prove this estimate, let B be the unit ball and set

$$M(\rho) = \max_{\rho B}\{u\}, \quad m(\rho) = \min_{\rho B}\{u\}, \quad 0 < \rho < 1.$$

Fix $0 < \rho < 1$ and set $M = M(\rho)$, $m = m(\rho)$, $M' = M(\rho/2)$, $m' = m(\rho/2)$. Then, $M - u$ and $u - m$ are non-negative solutions in ρB. Thus, by the Harnack inequality of Theorem 2.1.1,

$$\max_{(\rho/2)B} \{M - u\} \le C \inf_{(\rho/2)B} \{M - u\},$$

$$\max_{(\rho/2)B} \{u - m\} \le C \inf_{(\rho/2)B} \{u - m\},$$

that is,

$$M - m' \le C(M - M'), \quad M' - m \le C(m' - m).$$

Hence

$$M - m + M' - m' \le C(M - m - M' + m'),$$

that is,

$$M' - m' \le \frac{C - 1}{C + 1}(M - m).$$

Thus the oscillation $\omega(\rho) = M(\rho) - m(\rho)$ satisfies

$$\omega(2^{-\ell}) \le 2^{-\alpha\ell}\omega(1)$$

with $\alpha = \log[(C + 1)/(C - 1)]$. It follows that

$$\omega(\rho) \le 2^\alpha \rho^\alpha \omega(1).$$

After scaling, this yields the following lemma.

Lemma 2.3.2 *There exists* $\alpha = \alpha(n, \lambda)$ *such that any solution* u *of* $Lu = 0$ *in a ball* B *satisfies*

$$\forall \rho \in (0, 1), \quad \sup_{x,y \in \rho B} \{|u(x) - u(y)|\} \le 2^{1+\alpha} \rho^\alpha \sup_B\{u\}.$$

We can now finish the proof of (2.3.3). Assume that B is the unit ball. Fix $\delta \in (0, 1)$. For $x, y \in \delta B$, consider two cases. If $|x - y| \ge (1 - \delta)$, then

$$|u(x) - u(y)| \le 2\sup_B\{u\} \le 2(1 - \delta)^{-\alpha}|x - y|^\alpha \sup_B\{u\}.$$

If $|x - y| \le (1 - \delta)$ then the ball B' of radius $(1 - \delta)$ and center $(x + y)/2$ contains both x and y and is contained in B. Moreover, x, y are contained in $(|x - y|/(1 - \delta))B'$. Applying Lemma 2.3.2 in B' yields

$$|u(x) - u(y)| \le 2^{1+\alpha}(1 - \delta)^{-\alpha}|x - y|^\alpha \sup_B\{u\}.$$

This proves (2.3.3) and finishes the proof of Theorem 2.1.1.

Chapter 3

Sobolev inequalities on manifolds

3.1 Introduction

3.1.1 Notation concerning Riemannian manifolds

We want now to replace the Euclidean space $\mathbb{R}^n$ by a Riemannian manifold M and consider the possibility of having some kind of Sobolev inequalities. This brings in a whole new point of view. On Euclidean space, we could only discuss whether inequalities were true or not. In the more general setting of Riemannian manifolds, we can investigate the relations between various functional inequalities and the relations between these functional inequalities and the geometry of the manifold. We can search for necessary and/or sufficient conditions for a given Sobolev-type inequality to hold true. This leads to a better understanding of what information about M is encoded in various Sobolev-type inequalities.

Sobolev inequalities are useful when developing analysis on Riemannian manifolds, even more so than on Euclidean space, because other tools such as Fourier analysis are not available any more. This is particularly true when one studies large scale behavior of solutions of partial differential equations such as the Laplace and heat equations.

In the sequel, we will focus on complete, non-compact Riemannian manifolds. For compact manifolds, local Euclidean-type Sobolev inequalities are always satisfied and the interesting questions have to do with controlling the constants arising in these inequalities in geometric terms. We refer the interested reader to [5, 39, 40] where this is discussed at length.

Let us briefly introduce notation concerning Riemannian manifolds. Let M be a Riemannian manifold of dimension n with tangent space TM and co-tangent space T^*M. The tangent space TM is the union of the tangent spaces T_x where $x \in M$. T_x^* is the dual of the n-dimensional linear space T_x and T^*M is the union of these spaces. Smooth sections of TM are called vector fields and smooth sections of T^*M are called forms (1-forms). There is a natural pairing $TM \times T^*M \ni (\xi, \eta) \mapsto \eta(\xi) \in \mathbb{R}$ induced by the natural pairing of T_x and T_x^*, $x \in M$. It is a basic fact that vector fields can equivalently be defined as derivations, i.e., maps $\xi : \mathcal{C}^\infty(M) \to \mathcal{C}^\infty(M)$ such that $\xi(fg) = f\xi g + g\xi f$ for all $f, g \in \mathcal{C}^\infty(M)$. If f is a smooth function on M, the relation between its derivative df which is a form and ξf where ξ is a vector field is given by

$$df(\xi) = \xi f.$$

Because M is a Riemannian manifold, we are given on each T_x a scalar product $\langle \cdot, \cdot \rangle_x$. If $f \in \mathcal{C}^\infty(M)$, its gradient is defined as the unique vector field ∇f such that

$$\forall\, x \in M, \;\; \forall \xi \in TM, \;\; \langle \nabla f(x), \xi(x) \rangle_x = df(\xi)(x).$$

There is a canonical distance function associated to the Riemannian structure of M. We will denote it by $(x, y) \mapsto d(x, y)$. It can be defined as the shortest length of all piecewise $\mathcal{C}^1$ curves from x to y. The topology of (M, d) as a metric space is the same as that of M as a manifold. See, e.g., [13, §1.6]. There is also a canonical Riemannian measure on M which we denote by either dx or μ depending on which is more convenient. See, e.g., [13, §3.3]. We denote by $V(x, t)$ the volume of the ball of radius $t > 0$ around $x \in M$, i.e.,

$$V(x, t) = \mu(B(x, t)).$$

Thus $V(x, t)$ describes the volume growth of M.

The divergence $\operatorname{div}\xi$ of a vector field ξ is defined as the unique smooth function on M such that

$$\forall\, f \in \mathcal{C}_0^\infty(M), \;\; \int_M f \operatorname{div}\xi \, d\mu = -\int_M df(\xi) d\mu.$$

The Laplace–Beltrami operator Δ on M is the second order differential operator defined by

$$\forall\, f \in \mathcal{C}_0^\infty(M), \;\; \Delta f = -\operatorname{div}(\nabla f).$$

Note that, with this definition,

$$\forall f, g \in \mathcal{C}_0^\infty(M), \;\; \int_M f \Delta g \, d\mu = \int_M \langle \nabla f, \nabla g \rangle d\mu,$$

so that, in particular,

$$\int_M f \Delta f \, d\mu = \int_M \langle \nabla f, \nabla f \rangle d\mu \geq 0.$$

All the objects introduced above can of course be computed in local coordinates. See, e.g., [12, 13].

We will always work under the assumption that M is complete. Although this could a priori be interpreted to have various metric or geometric meanings (e.g., geodesically complete), the different interpretations turn out to be equivalent. Thus, M is complete means that (M, d) is a complete metric space. In particular, all bounded closed sets are compact. See [13, §1.7].

3.1.2 Isoperimetry

On a Riemannian manifold, any smooth $(n - 1)$-submanifold (i.e., hypersurface of co-dimension 1) inherits a Riemannian measure which we will denote by μ_{n-1}. The isoperimetric problem on M asks for the maximal volume that can be enclosed in a hypersurface of prescribed $(n-1)$-volume and for a description of the extremal sets, if they exist.

The first part of this problem can be interpreted as the search for some function Φ (depending on M) such that

$$\Phi(\mu_n(\Omega)) \leq \mu_{n-1}(\partial\Omega)$$

for all bounded sets $\Omega \subset M$ with smooth boundary $\partial\Omega$.

Solving the second part of the isoperimetric problem of course yields such an inequality. For instance, if M is n-dimensional Euclidean space, balls are extremal sets for the isoperimetric problem and this leads to the optimal function Φ given by

$$\Phi_{\mathbb{R}^n}(t) = \frac{\omega_{n-1}}{\Omega_n^{1-1/n}} t^{1-1/n}.$$

In view of this fundamental example, it is natural to consider the possibility that a Riemannian manifold M satisfies

$$\mu_n(\Omega)^{1-1/\nu} \leq C(M,\nu)\mu_{n-1}(\partial\Omega) \qquad (3.1.1)$$

for some constant $\nu > 0$ and $C(M,\nu) > 0$. Note that this could possibly be satisfied for a number of different values of ν and that the set of possible values of ν is either empty or an interval.

Theorem 3.1.1 *Assume that M satisfies* (3.1.1) *for some positive finite ν and $C(M,\nu)$. Then*

$$V(x,r) \geq c(M,\nu)r^\nu$$

where $c(M,\nu) = [\nu C(M,\nu)]^{-\nu}$. In particular, if M is n-dimensional and satisfies (3.1.1) *then $\nu \geq n$.*

The proof is straightforward if one observes that $\partial_r V(x,r)$ is the $(n-1)$-dimensional Riemannian volume of the boundary of $B(x,r)$. Indeed, we then have

$$V(x,r)^{1-1/\nu} \leq C(M,\nu)\partial_r V(x,r).$$

That is

$$\partial_r \left[V(x,r)^{1/\nu} \right] \geq [\nu C(M,\nu)]^{-1}.$$

This obviously yields

$$V(x,r) \geq [\nu C(M,\nu)]^{-\nu} r^{\nu}.$$

However, this proof is not quite complete because the boundary of a ball need not be a smooth $(n-1)$-submanifold (for large radius r). Here, we will ignore this difficulty and refer the reader to [13, §3.3,3.5] for details justifying the computation above. A different proof of this lemma, avoiding this difficulty, will be given later on (see Theorem 3.1.5 below).

In Euclidean space, we noticed the formal equivalence of the isoperimetric inequality with the Sobolev inequality $\|f\|_{n/(n-1)} \leq C_n \|\nabla f\|_1$. The argument, based on the co-area formula (see, e.g., [13, Theorems 3.13 and 6.3]), works as well on any Riemannian manifold. This proves the following important result.

Theorem 3.1.2 *A manifold M satisfies the inequality* (3.1.1) *for some positive ν and $C(M,\nu)$ if and only if it satisfies the inequality*

$$\forall f \in \mathcal{C}_0^\infty(M), \quad \|f\|_{\nu/(\nu-1)} \leq C(M,\nu)\|\nabla f\|_1.$$

Let us fix p,ν such that $1 \leq p < \nu$. We say that a Riemannian manifold M satisfies an (L^p,ν)-Sobolev inequality if there exists a constant $C(M,p,\nu)$ such that

$$\forall f \in \mathcal{C}_0^\infty(M), \quad \|f\|_{p\nu/(\nu-p)} \leq C(M,p,\nu)\|\nabla f\|_p.$$

Thus, Theorem 3.1.2 can be interpreted as saying that a manifold M satisfies an (L^1,ν)-Sobolev inequality if and only if it satisfies (3.1.1). The next result shows that the strength of an (L^p,ν)-Sobolev inequality decreases as p increases.

Theorem 3.1.3 *If M satisfies an (L^p,ν)-Sobolev inequality then it satisfies an (L^q,ν)-Sobolev inequality for all $p \leq q < \nu$.*

Apply the (L^p,ν) inequality to $|f|^\gamma$ for some $\gamma > 1$ to be fixed later. Thus

$$\|f\|_{\gamma p\nu/(\nu-p)}^\gamma \leq \gamma C(M,p,\nu) \left(\int_M |f|^{p(\gamma-1)} |\nabla f|^p d\mu \right)^{1/p}.$$

Now, apply the Hölder inequality

$$\int hg d\mu \leq \|h\|_{q/(q-p)} \|g\|_{q/p}$$

with $h = |f|^{p(\gamma-1)}$ and $g = |\nabla f|^p$. This yields

$$\left(\int_M |f|^{p(\gamma-1)} |\nabla f|^p d\mu \right)^{1/p}$$

$$\leq \left(\int |f|^{pq(\gamma-1)/(q-p)} d\mu \right)^{1/p-1/q} \left(\int |\nabla f|^q d\mu \right)^{1/q}.$$

Hence,

$$\|f\|_{\gamma p\nu/(\nu-p)}^{\gamma} \leq \gamma C(M,p,\nu) \left(\int |f|^{pq(\gamma-1)/(q-p)} d\mu \right)^{1/p-1/q} \left(\int |\nabla f|^q d\mu \right)^{1/q}.$$

Picking $\gamma = q(n-p)/p(n-q)$ and computing $\gamma - 1 = n(q-p)/p(n-q)$ yields

$$\|f\|_{qn/(n-q)} \leq \frac{q(n-p)}{p(n-q)} C(M,p,\nu) \|\nabla f\|_q$$

as desired.

3.1.3 Sobolev inequalities and volume growth

The next lemma introduces a family of (a priori weaker) inequalities that can be deduced from a Sobolev inequality. Our aim in this section is to show that weak forms of Sobolev inequalities are sufficient to imply a lower bound for the volume growth of M.

Lemma 3.1.4 *If M satisfies an (L^p, ν)-Sobolev inequality*

$$\forall f \in \mathcal{C}_0^\infty(M), \quad \|f\|_{p\nu/(\nu-p)} \leq C(M,p,\nu) \|\nabla f\|_p$$

then it also satisfies

$$\forall f \in \mathcal{C}_0^\infty(M), \quad \|f\|_r \leq (C(M,p,\nu) \|\nabla f\|_p)^\theta \|f\|_s^{1-\theta}$$

for all $0 < r, s < \infty$ and $0 \leq \theta \leq 1$ such that

$$\frac{1}{r} = \theta \left(\frac{1}{\nu} - \frac{1}{p} \right) + (1-\theta) \frac{1}{s}.$$

Set $q = \nu p/(\nu-p)$. Then $1/r = \theta/q + (1-\theta)/s$ and, by the Hölder inequality,

$$\|f\|_r \leq \|f\|_q^\theta \|f\|_s^{1-\theta}.$$

This yields the desired inequality. The possible range for s is actually $(0, \nu p/(\nu-p))$ and the range of r is $(s, \nu p/(\nu-p))$. Note that when $\theta = 0$ one must have $r = s$ and the conclusion of the lemma is trivial.

The following result generalizes Theorem 3.1.1.

Theorem 3.1.5 *Assume that the inequality*

$$\forall f \in \mathcal{C}_0^\infty(M), \quad \|f\|_r \leq (C\|\nabla f\|_p)^\theta \|f\|_s^{1-\theta}$$

is satisfied for some r, s, θ *with* $0 < s \leq r \leq \infty$ *and* $0 < \theta \leq 1$. *Assume also that*

$$\frac{\theta}{p} + \frac{1-\theta}{s} - \frac{1}{r} > 0.$$

Then

$$V(x,t) \geq ct^\nu$$

with ν *defined by*

$$\frac{\theta}{\nu} = \frac{\theta}{p} + \frac{1-\theta}{s} - \frac{1}{r} \tag{3.1.2}$$

and the constant c *given by*

$$c = 2^{-\nu^2/\theta r - \nu/\theta} C^{-\nu}.$$

In particular, if M *satisfies an* (L^p, ν)-*Sobolev inequality for some* p, ν *with* $1 \leq p < \nu$, *then* $n \geq \nu$ *and the volume growth function* $V(x,t)$ *satisfies*

$$\inf_{\substack{x \in M \\ t > 0}} \{t^\nu V(x,t)\} > 0.$$

We will use the fact that the distance function has gradient bounded by 1 almost everywhere. Indeed, for any fixed $x \in M$ and $t > 0$, consider the function

$$f(y) = \max\{t - d(x,y)), 0\}.$$

Then

$$\begin{aligned}
\|f\|_r &\geq (t/2)V(x, t/2)^{1/r} \\
\|f\|_s &\leq tV(x,t)^{1/s} \\
\|\nabla f\|_p &\leq V(x,t)^{1/p}.
\end{aligned}$$

Hence

$$V(x,t)^{\theta/p + (1-\theta)/s} \geq 2^{-1}(t/C)^\theta V(x, t/2)^{1/r}.$$

If we could ignore the fact that we have the volume of the ball of radius $t/2$ instead of t on the right-hand side of this inequality, we would get $V(x,t) \geq ct^\nu$ with ν given by $\theta/\nu = \theta/p + (1-\theta)/s - 1/r$. Thus, define ν by

$$\frac{\theta}{\nu} = \frac{\theta}{p} + \frac{1-\theta}{s} - \frac{1}{r}$$

and write the last inequality in the form

$$V(x,t) \geq (2C^\theta)^{-r\nu/(\nu+\theta r)} t^{\theta r\nu/(\nu+\theta r)} V(x, t/2)^{\nu/(\nu+\theta r)}.$$

It follows that

$$V(x,t) \geq (2C^\theta)^{-r\sum_1^i a^j} t^{\theta r \sum_1^i a^j} 2^{-\theta r \sum_1^i (j-1)a^j} V(x,t/2^i)^{a^i} \qquad (3.1.3)$$

with $a = \nu/(\nu + \theta r)$. Observe that $a < 1$ as long as $\theta \neq 0$. Moreover, in this case,

$$\sum_1^\infty a^j = a(1-a)^{-1} = \frac{\nu}{\theta r}, \quad \sum_1^\infty (j-1)a^j = a^2(1-a)^{-2} = \frac{\nu^2}{\theta^2 r^2}.$$

Furthermore, $\lim_{t \to 0} t^{-n} V(x,t) = \Omega_n$. Hence for i large enough,

$$\liminf_{i \to \infty} V(x,t/2^i)^{a_i} \geq \lim_{i \to \infty} [\Omega_n t^n / 2^{in+1}]^{a^i} = \lim_{i \to \infty} 2^{-ina^i} = 1.$$

Letting i tend to infinity in (3.1.3), we obtain

$$V(x,r) \geq 2^{-\nu^2/\theta r} (2C^\theta)^{-\nu/\theta} t^\nu.$$

Note that this proof uses the (Riemannian) fact that $\lim_{t\to 0} t^{-n} V(x,t) = \Omega_n$. Later, we will give another proof avoiding the use of this fact.

It is useful to illustrate Theorem 3.1.5 by a very simple case showing how the volume parameter ν abstractly defined by (3.1.2) is computed from θ, r, s, p. Take $M = \mathbb{R}$. Then we have the calculus inequality,

$$\forall f \in \mathcal{C}_0^\infty(\mathbb{R}), \quad \|f\|_\infty \leq (1/2)\|\nabla f\|_1.$$

Theorem 3.1.5 applies with $\theta = 1$, $r = \infty$, $p = 1$ (s is irrelevant when $\theta = 1$) and yields $\nu = 1$ and $V(t) \geq t$ (note that this is off only by a factor of 2). Applying the calculus inequality above to $|f|^2$ and using Cauchy–Schwarz, we find that $\|f\|_\infty \leq (1/2)\|f\|_2^{1/2}\|\nabla f\|_2^{1/2}$ and thus $\|f\|_2^2 \leq (1/2)\|\nabla f\|_2^{1/2}\|f\|_2^{1/2}\|f\|_1$. That is,

$$\forall \mathcal{C}_0^\infty(\mathbb{R}), \quad \|f\|_2 \leq (1/2)^{1/2}\|\nabla f\|_2^{1/3}\|f\|_1^{2/3}.$$

This is the Nash inequality, in one dimension. In the next section, we will take a closer look at this type of inequality. For now, observe that we can apply Theorem 3.1.5 again, this time with $r = p = 2$, $s = 1$, $\theta = 1/3$. Then ν is defined by $1/(3\nu) = 1/6 + 2/3 - 1/2 = 2/6$, i.e., $\nu = 1$ again.

From the proof of Theorem 3.1.5, it is obvious that one can work under a weaker hypothesis and obtain the following statement.

Theorem 3.1.6 *Assume that the inequality*

$$\forall f \in \mathcal{C}_0^\infty(M), \quad \sup_{t>0}\left\{ t\mu(|f| > t)^{1/r} \right\} \leq (C\|\nabla f\|_p)^\theta \left(\|f\|_\infty \mu(\operatorname{supp}(f))^{1/s}\right)^{1-\theta}$$

is satisfied for some r, s, θ with $0 < s, r \leq \infty$ and $0 < \theta \leq 1$. Assume also that

$$\frac{\theta}{p} + \frac{1-\theta}{s} - \frac{1}{r} > 0.$$

Then

$$V(x, t) \geq ct^{\nu}$$

with $0 < \nu < \infty$ defined by

$$\frac{\theta}{\nu} = \frac{\theta}{p} + \frac{1-\theta}{s} - \frac{1}{r}$$

and the constant c given by

$$c = 2^{-\nu^2/\theta r - \nu/\theta} C^{-\nu}.$$

In this theorem, if $r = \infty$, the weak L^r-norm must be understood as $\|f\|_\infty$.

3.2   Weak and strong Sobolev inequalities

3.2.1   Examples of weak Sobolev inequalities

Any Sobolev inequality can be used to deduce a priori weaker inequalities through the use of the Hölder inequality and related inequalities. For instance, the Sobolev inequality

$$\|f\|_{2\nu/(\nu-2)} \leq C\|\nabla f\|_2$$

obviously implies the weak Sobolev inequality

$$\sup_{t>0} \left\{ t\mu(\{|f| > t\})^{(\nu-2)/2\nu} \right\} \leq C\|\nabla f\|_2.$$

We have also seen in Section 2.1.3 that the Sobolev inequality

$$\|f\|_{2\nu/(\nu-2)} \leq C\|\nabla f\|_2$$

implies

$$\|f\|_{2(1+2/\nu)} \leq (C\|\nabla f\|_2)^{\nu/(\nu+2)} \|f\|_2^{2/(\nu+2)}.$$

A different use of the Hölder inequality leads to

$$\|f\|_2 \leq (C\|\nabla f\|_2)^{\nu/(\nu+2)} \|f\|_1^{2/(\nu+2)}$$

which one often writes

$$\|f\|_2^{2(1+2/\nu)} \leq C^2 \|\nabla f\|_2^2 \|f\|_1^{4/\nu}.$$

This last inequality is called a Nash inequality. It first appeared in the 1958 paper of Nash [67] concerning the regularity of solutions of parabolic uniformly elliptic equations in divergence form in $\mathbb{R}^n$. Nash did not deduce this inequality from the Sobolev inequality. Instead, he gave the following direct proof.

Theorem 3.2.1 *There is a constant C_n such that any smooth compactly supported function f on $\mathbb{R}^n$ satisfies*

$$\|f\|_2^{2(1+2/n)} \le C_n^2 \|\nabla f\|_2^2 \|f\|_1^{4/n}.$$

Let $\hat{f}$ be the Fourier transform of f. Then

$$
\begin{aligned}
\|f\|_2^2 &= \|\hat{f}\|_2^2 = \int_{|\xi| \le R} |\hat{f}(\xi)|^2 d\xi + \int_{|\xi| > R} |\hat{f}(\xi)|^2 d\xi \\
&\le \Omega_n R^n \|\hat{f}\|_\infty^2 + R^{-2} \int |\hat{f}(\xi)|^2 |\xi|^2 d\xi \\
&\le \Omega_n R^n \|f\|_1^2 + (2\pi R)^{-2} \|\nabla f\|_2^2.
\end{aligned}
$$

Optimizing in R yields

$$\|f\|_2^2 \le (2+n)(\Omega_n/4\pi)^{2/(n+2)} \|\nabla f\|_2^{2n/(n+2)} \|f\|_1^{4/(n+2)},$$

which gives the desired inequality with

$$C_n = (2+n)^{1+2/n}(\Omega_n/4\pi)^{2/n}.$$

Nash's proof is given here to point out how different it is from the proofs of the corresponding Sobolev inequality that we have seen. The question naturally arises of whether or not this is yet another proof of the L^2-Sobolev inequality in $\mathbb{R}^n$. That is, can we easily deduce the Sobolev inequality

$$\|f\|_{2\nu/(\nu-2)} \le C \|\nabla f\|_2$$

from the Nash inequality

$$\|f\|_2^{2(1+2/\nu)} \le C^2 \|\nabla f\|_2^2 \|f\|_1^{4/\nu},$$

maybe with different constants C?

There is an interesting twist here because the Nash inequality of Theorem 3.2.1 is valid for all $n \ge 1$ whereas the corresponding Sobolev inequality is valid only for $n \ge 3$. In some sense, this means that, if Nash inequality implies Sobolev inequality, it cannot be in a completely obvious way.

In general, one can ask: do various weak forms of Sobolev inequality imply their stronger counterparts? The next few sections show that, somewhat surprisingly, the answer is yes. Moreover, this can be proved by some elementary and widely applicable arguments.

3.2.2 $(S_{r,s}^\theta)$-inequalities: the parameters q and ν

Let us fix the parameter p, $1 \le p < \infty$. We want to discuss functional inequalities of Sobolev type for smooth compactly supported functions under the hypothesis that we can control $\|\nabla f\|_p$. The weakest type of Sobolev inequality we have encountered so far is that of Theorem 3.1.6 which states that, for all $f \in \mathcal{C}_0^\infty(M)$,

$$\sup_{t>0} \left\{ t\mu(\{|f| > t\})^{1/r} \right\} \leq (C\|\nabla f\|_p)^\theta \left(\|f\|_\infty \mu(\mathrm{supp}(f))^{1/s} \right)^{1-\theta}.$$

We call this inequality $(S_{r,s}^{*,\theta})$. Recall that it has a slightly stronger version (see Theorem 3.1.5) which reads

$$\forall f \in \mathcal{C}_0^\infty(M), \quad \|f\|_r \leq (C\|\nabla f\|_p)^\theta \|f\|_s^{1-\theta}. \qquad (S_{r,s}^\theta)$$

In these inequalities, we can think of $0 < r, s \leq \infty$, $0 < \theta \leq 1$ as parameters and we would like to understand the meaning of these parameters.

For the time being, let us simply observe that $(S_{r,s}^1)$ (the parameter s plays no role when $\theta = 1$) is the classical Sobolev inequality

$$\|f\|_r \leq C\|\nabla f\|_p.$$

Similarly, $(S_{r,s}^{*,1})$ (again, the parameter s plays no role when $\theta = 1$) is the weak Sobolev inequality

$$\sup_{t>0} \left\{ t\mu(|f| > t)^{1/r} \right\} \leq C\|\nabla f\|_p.$$

Finally, when $p = 2$, $(S_{2,1}^\theta)$ is the Nash inequality

$$\|f\|_2^{2(1+(1-\theta)/\theta)} \leq C^2 \|\nabla f\|_2^2 \|f\|_1^{2(1-\theta)/\theta}.$$

Observe also that, by Theorem 3.1.6, each of these inequalities implies

$$\inf_{t>0}\{t^{-\nu}V(x,t)\} > 0$$

where the parameter ν is defined by $1/r = 1/p - 1/\nu$ for the first two inequalities (assuming $p < r$) and by $2/\nu = \theta/(1-\theta)$ for the Nash inequality.

It turns out that one of the keys to understanding these inequalities is to consider yet another parameter, call it $q = q(r, s, \theta)$, which belongs to $(-\infty, 0) \cup (0, +\infty) \cup \{\infty\}$ and which is defined by

$$\frac{1}{r} = \frac{\theta}{q} + \frac{1-\theta}{s}. \qquad (3.2.1)$$

Observe that q is related to the parameter ν of Theorems 3.1.5, 3.1.6 by

$$\frac{1}{q} = \frac{1}{p} - \frac{1}{\nu}. \qquad (3.2.2)$$

It is a fundamental difference between q and ν that q can be computed from (r, s, θ) without explicit reference to p whereas ν cannot.

We will prove below that, roughly speaking, any weak inequality $(S_{r_0,s_0}^{*,\theta_0})$ for some fixed r_0, s_0, θ_0 implies the full collection of strong inequalities

$$(S_{r,s}^{\theta}) \text{ for all } 0 < r, s \le \infty, \; 0 < \theta \le 1$$

with

$$q(r, s, \theta) = q(r_0, s_0, \theta_0).$$

More precisely, this is correct when

$$1/q(r_0, s_0, \theta_0) \le 0.$$

When $1/q(r_0, s_0, \theta_0) > 0$, one needs to assume also that $q \ge p$.

The proof will depend on elementary functional analysis arguments and repeated application of the given inequality to functions of the type $(f - u)_+ \wedge v$ with $u, v > 0$, $u \wedge v = \min\{u, v\}$, $u_+ = \max\{u, 0\}$, where f is a given fixed function in $\mathcal{C}_0^{\infty}(M)$.

Here, we consider the parameter p as fixed once and for all. Later, in Section 3.2.6, we will also consider what happens when p varies. Then the parameter ν will play a crucial role. Roughly speaking, any of the inequalities $(S_{r_0,s_0}^{*,\theta_0})$ for some fixed r_0, s_0, θ_0 and p_0 implies all the inequalities $(S_{r,s}^{\theta})$ for all $p \ge p_0$ and where the parameters r, s, θ, p satisfy *both* (3.2.1) and (3.2.2).

3.2.3   The case $0 < q < \infty$

In this section the number p, $1 \le p < \infty$, is fixed and all Sobolev-type inequalities are relative to $\|\nabla f\|_p$. The main result of this section is described in the following theorem.

Theorem 3.2.2 *Assume that $(S_{r_0,s_0}^{*,\theta_0})$ is satisfied for some $0 < r_0, s_0 \le \infty$ and $0 < \theta_0 \le 1$ and let $q = q(r_0, s_0, \theta_0)$ be defined as in (3.2.1). Assume that $p \le q < \infty$. Then all the inequalities $(S_{r,s}^{\theta})$ with $0 < r, s \le \infty$, $0 < \theta \le 1$ and $q(r, s, \theta) = q$ are also satisfied. In particular, there exists a finite constant A such that*

$$\forall f \in \mathcal{C}_0^{\infty}(M), \quad \|f\|_q \le AC\|\nabla f\|_p$$

where C is the constant appearing in $(S_{r_0,s_0}^{*,\theta_0})$. That is, M satisfies an (L^p, ν)-Sobolev inequality with ν defined by $1/q = 1/p - 1/\nu$.

Fix a non-negative function $f \in \mathcal{C}_0^{\infty}(M)$ and set

$$f_{\rho,k} = (f - \rho^k)_+ \wedge \rho^k(\rho - 1) \tag{3.2.3}$$

for any $\rho > 1$ and $k \in \mathbb{Z}$. This function has the following properties. Its support is contained in $\{f \ge \rho^k\}$. Moreover,

$$\{f_{\rho,k} \ge (\rho - 1)\rho^k\} = \{f \ge \rho^{k+1}\}. \tag{3.2.4}$$

Finally, $f_{\rho,k}$ has the same "profile" as f on $\{\rho^k < f < \rho^{k+1}\}$ and is flat outside this set. In particular,

$$|\nabla f_{\rho,k}| \leq |\nabla f|. \tag{3.2.5}$$

By (3.2.4), (3.2.5) and our hypothesis applied to $f_{\rho,k}$, we have

$$(\rho - 1)\rho^k \mu(f \geq \rho^{k+1})^{1/r_0} \leq (C\|\nabla f\|_p)^{\theta_0} \left((\rho - 1)\rho^k \mu(f \geq \rho^k)^{1/s_0}\right)^{1-\theta_0}.$$

Let us set

$$N(f) = \sup_k \left\{\rho^k \mu(f \geq \rho^k)^{1/q}\right\}.$$

Using the definition of q we then get

$$\mu(f \geq \rho^{k+1})^{1/r_0}$$

$$\leq \quad \rho^{-k(\theta_0 + (1-\theta_0)q/s_0)} \left(\frac{C\|\nabla f\|_p}{\rho - 1}\right)^{\theta_0} N(f)^{q(1-\theta_0)/s_0}$$

$$\leq \quad \rho^{-kq/r_0} \left(\frac{C\|\nabla f\|_p}{\rho - 1}\right)^{\theta_0} N(f)^{q(1-\theta_0)/s_0}.$$

Thus

$$N(f)^{q/r_0} \leq \rho^{q/r_0} \left(\frac{C\|\nabla f\|_p}{\rho - 1}\right)^{\theta_0} N(f)^{q(1-\theta_0)/s_0}.$$

Simplifying and using the definition of q again yields

$$N(f) \leq \frac{\rho^{q/r_0}}{\rho - 1} C\|\nabla f\|_p \tag{3.2.6}$$

and thus

$$\sup_{t>0}\{t\mu(f \geq t)^{1/q}\} \leq \rho N(f) \leq \frac{\rho^{1+q/r_0}}{\rho - 1} C\|\nabla f\|_p.$$

Setting $\rho = 1 + r_0\theta_0/q$, we get

$$\sup_{t>0}\{t\mu(f \geq t)^{1/q}\} \leq e(1 + q/(r_0\theta_0))C\|\nabla f\|_p. \tag{3.2.7}$$

This is the weak form (in the sense of weak L^r spaces) of the desired (L^p, ν)-Sobolev inequality

$$\|f\|_{p\nu/(\nu-p)} \leq AC\|\nabla f\|_p$$

where ν is given by $1/q = 1/p - 1/\nu$. Thus, to finish the proof of Theorem 3.2.2, it suffices to prove the following lemma.

Lemma 3.2.3 *Assume that for some* $1 \leq p < q < \infty$,

$$\forall f \in \mathcal{C}_0^\infty(M), \ \sup_{t>0}\{t\mu(f \geq t)^{1/q}\} \leq C_1\|\nabla f\|_p.$$

Then

$$\forall f \in \mathcal{C}_0^\infty(M), \ \|f\|_q \leq 2(1 + q)C_1\|\nabla f\|_p.$$

For this lemma, we need to improve (3.2.5) and observe that, actually,

$$|\nabla f_k| \leq |\nabla f| \mathbf{1}_{\{\rho^k < f \leq \rho^{k+1}\}}. \tag{3.2.8}$$

Now, applying the hypothesis to $f_{\rho,k}$ yields

$$(\rho - 1)\rho^k \mu(f \geq \rho^{k+1})^{1/q} \leq C_1 \|\nabla f_k\|_p.$$

Raise this inequality to the power q and sum over $k \in \mathbb{Z}$ to get

$$(\rho - 1)^q \sum_k \rho^{kq} \mu(f \geq \rho^{k+1}) \leq C_1^q \sum_k \left(\int_{\{\rho^k < f \leq \rho^{k+1}\}} |\nabla f|^p \right)^{q/p}. \tag{3.2.9}$$

As $q/p \geq 1$, we have $\sum a_k^{q/p} \leq (\sum a_k)^{q/p}$ for any sequence (a_k) of non-negative reals. Also

$$\begin{aligned} \sum \rho^{kq} \mu(f \geq \rho^{k+1}) &\geq [\rho^q(\rho^q - 1)]^{-1} q \sum \int_{\rho^{k+1}}^{\rho^{k+2}} t^{q-1} \mu(f \geq t) dt \\ &= [\rho^q(\rho^q - 1)]^{-1} q \int_0^\infty t^{q-1} \mu(f \geq t) dt \\ &= [\rho^q(\rho^q - 1)]^{-1} \|f\|_q^q. \end{aligned}$$

Hence (3.2.9) yields

$$\|f\|_q \leq \frac{\rho(\rho^q - 1)^{1/q}}{(\rho - 1)} C_1 \|\nabla f\|_p.$$

Using $(\rho^q - 1)^{1/q} \leq \rho$ and picking $\rho = 1 + 1/q \leq 2$ yields the desired inequality.

Corollary 3.2.4 *For $\nu > 2$, the Nash-type inequality*

$$\forall f \in \mathcal{C}_0^\infty(M), \quad \|f\|_2^{2(1+2/\nu)} \leq C^2 \|\nabla f\|_2^2 \|f\|_1^{4/\nu}$$

implies the (L^2, ν)-Sobolev inequality

$$\forall f \in \mathcal{C}_0^\infty(M), \quad \|f\|_{2\nu/(\nu-2)} \leq A_\nu C \|\nabla f\|_2$$

for some constant A_ν.

Indeed, the postulated Nash-type inequality is exactly $(S_{2,1}^{\nu/(\nu+2)})$ raised to the power $2(\nu+2)/\nu$. Furthermore, $q = q(2, 1, \nu/(\nu+2))$ is equal to $2\nu/(\nu-2)$ which is finite and larger than 2 if ν is larger than 2. Thus we can apply Theorem 3.2.2 which yields the desired Sobolev inequality.

3.2.4   The case $q = \infty$

As in the previous section, the number p, $1 \leq p < \infty$, is fixed and all Sobolev-type inequalities are relative to $\|\nabla f\|_p$.

Theorem 3.2.5 *Assume that* $(S_{r_0,s_0}^{*,1-s_0/r_0})$ *is satisfied for some* $0 < s_0 < r_0 < \infty$ *(this has* $q(r_0, s_0, 1 - s_0/r_0) = \infty$ *for* q *given at (3.2.1)). Then all the inequalities* $(S_{r,s}^{1-s/r})$ *with* $0 < s < r < \infty$ *are satisfied.*

Fix a non-negative $f \in \mathcal{C}_0^\infty(M)$ and apply the hypothesis to $f_{\rho,k}$ defined at (3.2.3). Thanks to (3.2.4), (3.2.5), this gives

$$\rho^{kr_0}\mu(f \geq \rho^{k+1}) \leq \left(\frac{C\|\nabla f\|_p}{\rho - 1}\right)^{r_0 - s_0} \rho^{ks_0}\mu(f \geq \rho^k). \qquad (3.2.10)$$

Fix $0 < s < r < \infty$ and observe that it is enough to prove the desired inequality for some set S of pairs (r, s), $0 < s < r < \infty$ such that $\sup\{s : (r, s) \in S\} = \infty$. Indeed, the Hölder inequality shows that, if $r \geq s$, then $(S_{r,s}^\theta)$ implies $(S_{r',s'}^{\theta'})$ for all $r' \geq s'$ such that $r' \leq r$ and $s' \leq s$. For instance, it is enough to prove $(S_{r,s}^{1-s/r})$ for $(r, s) \in S = \{(r_0 + t, s_0 + t) : t > 0\}$. Now, fix any $t > 0$ and set $r = r_0 + t$, $s = s_0 + t$. Multiply (3.2.10) by q^{kt} to obtain

$$\rho^{kr}\mu(f \geq \rho^{k+1}) \leq \left(\frac{C\|\nabla f\|_p}{\rho - 1}\right)^{r_0 - s_0} \rho^{ks}\mu(f \geq \rho^k).$$

Summing over all k, we get

$$\sum_k \rho^{kr}\mu(f \geq \rho^{k+1}) \leq \left(\frac{C\|\nabla f\|_p}{\rho - 1}\right)^{r-s} \sum_k \rho^{ks}\mu(f \geq \rho^k).$$

But

$$\frac{1}{\rho^r(\rho^r - 1)}\|f\|_r^r \leq \frac{1}{\rho^r(\rho^r - 1)}\sum_k r \int_{\rho^{k+1}}^{\rho^{k+2}} t^{s-1}\mu(f \geq t)dt \leq \rho^{kr}\mu(f \geq \rho^{k+1})$$

and, similarly,

$$\sum_k \rho^{ks}\mu(f \geq \rho^k) \leq \frac{\rho^s}{\rho^s - 1}\|f\|_s^s.$$

Hence

$$\|f\|_r^r \leq \frac{\rho^{r+s}(\rho^r - 1)}{(\rho^s - 1)(\rho - 1)^{r-s}}(C\|\nabla f\|_p)^{r-s}\|f\|_s^s$$

for all $t > 1$ and $r = r_0 + t$, $s = s_0 + t$. For any fixed $\rho > 1$, this gives the desired inequality for all $0 < s < r < \infty$. This proves Theorem 3.2.5.

We now want to obtain a result that complements Theorem 3.2.5 under the same hypothesis. Observe that we cannot take $r = \infty$ in Theorem 3.2.5.

Actually, the case of $\mathbb{R}^n$ shows that we cannot hope to have an inequality of the form

$$\|f\|_\infty \le (C\|\nabla f\|_p)^\theta \|f\|_s^{1-\theta}$$

when $q = \infty$. What we can hope for is some form of local exponential integrability. This can be obtained by looking more closely at the proof given above. By Theorem 3.2.5, we can now assume without loss of generality that $(S_{2,1}^{1/2})$ is satisfied. Repeating the argument above and using $\gamma(x-1) \le x^\gamma - 1 \le \gamma(x-1)x^{\gamma-1}$ which is valid for $\gamma \ge 1$, we obtain that there exists a constant C such that

$$\|f\|_r \le 2^{1/r}\rho^3(\rho-1)^{-1+s/r}(C\|\nabla f\|_p)^{1-s/r}\|f\|_s^{s/r}$$

for all $t > 0$, $r = 2 + t$, $s = 1 + t$. Let Ω be the support of f. As $1 \le s < r$, we have $\|f\|_s \le \mu(\Omega)^{1/s-1/r}\|f\|_r$. Hence, for any $r \ge 2$ and $s = r - 1$

$$\|f\|_r \le 2^{1/r}\rho^3(\rho-1)^{-1+s/r}(C\|\nabla f\|_p)^{1-s/r}\mu(\Omega)^{1-s/r}\|f\|_r^{s/r}$$

which gives (recall that $r - s = 1$)

$$\|f\|_r \le 2\rho^{3r}(\rho-1)^{-1}C\|\nabla f\|_p\mu(\Omega)^{1/r}$$

for all $r \ge 2$. Picking $\rho = 1 + 1/r$ yields

$$\|f\|_r \le 54\,r\,C\|\nabla f\|_p\mu(\Omega)^{1/r}.$$

Hence we have obtained the following result.

Theorem 3.2.6 *Under the hypothesis of Theorem 3.2.5 there exists a constant A_1 such that for all bounded sets $\Omega \subset M$ and all $r \ge 1$,*

$$\forall f \in \mathcal{C}_0^\infty(\Omega), \quad \|f\|_r \le A_1(1+r)\mu(\Omega)^{1/r}\|\nabla f\|_p.$$

Moreover, there exist two constants $\alpha > 0$ and A_2 such that for all bounded sets $\Omega \subset M$,

$$\forall f \in \mathcal{C}_0^\infty(\Omega), \quad \int_\Omega e^{\alpha|f|/\|\nabla f\|_p}d\mu \le A_2\mu(\Omega).$$

The second inequality stated above is an easy consequence of the first. Note that this is not the sharpest result that can be obtained. Indeed, as in the case of $\mathbb{R}^n$, one can show that $(S_{r_0,s_0}^{*,1-s_0/r_0})$ for some fixed $0 < s_0 < r_0 < \infty$ implies the stronger integrability

$$\forall f \in \mathcal{C}_0^\infty(\Omega), \quad \int_\Omega e^{(\alpha'|f|/\|\nabla f\|_p)^{p/(p-1)}}d\mu \le A'\mu(\Omega)$$

for some constants $A', \alpha' > 0$. See [6] for a proof of this sharper result using the same ideas. Interestingly enough, in order to obtain the sharper result one apparently needs to use Lorentz spaces instead of mere L^r-spaces.

3.2.5   The case $-\infty < q < 0$

As in the previous two sections, the number p, $1 \le p < \infty$, is fixed and all Sobolev-type inequalities are relative to $\|\nabla f\|_p$.

Theorem 3.2.7 *Assume that $(S^{*,\theta_0}_{r_0,s_0})$ is satisfied for some $0 < r_0, s_0 \le \infty$ and $0 < \theta_0 \le 1$ and let $q = q(r_0, s_0, \theta_0)$ be defined as in (3.2.1). Assume that $-\infty < q < 0$ (this forces $s_0 < r_0$). Then all the inequalities $(S^{\theta}_{r,s})$ with $0 < s < r \le \infty$, $0 < \theta \le 1$ and $q(r,s,\theta) = q$ are also satisfied. In particular, there exists a finite constant A such that*

$$\forall f \in \mathcal{C}_0^\infty(M), \quad \|f\|_\infty \le A(C\|\nabla f\|_p)^{1/(1-s/q)}\|f\|_s^{1/(1-q/s)},$$

for all $0 < s < \infty$ (recall that $q < 0$). Here C is the constant appearing in $(S^{*,\theta_0}_{r_0,s_0})$.

Fix $f \in \mathcal{C}_0^\infty(M)$, $f \ge 0$ and $\|f\|_\infty \ne 0$. Fix also $\epsilon > 0$ small enough and $\rho > 1$. Define the functions $f_{\rho,k}$ by setting

$$f_{\rho,k} = (f - (\|f\|_\infty - \epsilon - \rho^k))_+ \wedge \rho^{k-1}(\rho - 1)$$

for all $k \le k(f)$ where $k(f)$ is the largest integer k such that $\rho^k < \|f\|_\infty$. Note that $f_{\rho,k}$ is compactly supported if $0 < \epsilon \le \|f\|_\infty - \rho^{k(f)}$ and that $|\nabla f_{\rho,k}| \le |\nabla f|$ for all $k \le k(f)$. Set

$$\lambda_k = \|f\|_\infty - \epsilon - \rho^k.$$

Observe that $f_{\rho,k}$ has support in $\{f \ge \lambda_k\}$ and that

$$\{f_{\rho,k} \ge \rho^{k-1}(\rho - 1)\} = \{f \ge \lambda_{k-1}\}.$$

Applying $(S^{*,\theta}_{r_0,s_0})$ to $f_{\rho,k}$, we obtain

$$\rho^{k-1}\mu(f \ge \lambda_{k-1})^{1/r_0} \le \left(\frac{C\|\nabla f\|_p}{\rho - 1}\right)^{\theta_0} \rho^{(k-1)(1-\theta_0)}\mu(f \ge \lambda_k)^{(1-\theta_0)/s_0}.$$

Multiply this inequality by $\rho^{\delta(k-1)}$ and rearrange to obtain

$$\left(\rho^{r_0(k-1)(1+\delta)}\mu(f \ge \lambda_{k-1})\right)^{1/r_0}$$
$$\le \left(\frac{C\|\nabla f\|_p}{\rho - 1}\right)^{\theta_0} \left[\rho^{s_0(k-1)(1+\delta/(1-\theta_0))}\mu(f \ge \lambda_k)\right]^{(1-\theta_0)/s_0}.$$

Now, choose δ so that $r_0(1 + \delta) = s_0(1 + \delta/(1 - \theta_0))$. It turns out that this is equivalent to $r_0(1 + \delta) = q$. Setting $a_k = \rho^{qk}\mu(f \ge \lambda_k)$ yields

$$a_{k-1}^{1/r_0} \le \rho^{-q(1-\theta_0)/s_0} \left(\frac{C\|\nabla f\|_p}{\rho - 1}\right)^{\theta_0} a_k^{(1-\theta_0)/s_0} \qquad (3.2.11)$$

for all $k \leq k(f)$. Observe that $a_k > 0$ for all $k \leq k(f)$ and that

$$\lim_{k \to -\infty} a_k = \infty$$

because $\mu(f \geq \|f\|_\infty - \epsilon) > 0$ and $q < 0$. This is actually the reason why the parameter ϵ was introduced in the computation above. Because of these observations, it is clear that

$$a = \inf_{k \leq k(f)} a_k$$

is positive. It follows that (3.2.11) implies

$$a^{1/r_0} \leq \rho^{-q(1-\theta_0)/s_0} \left(\frac{C\|\nabla f\|_p}{\rho - 1} \right)^{\theta_0} a^{(1-\theta_0)/s_0},$$

that is,

$$a = \inf_{k \leq k(f)} \left\{ \rho^{qk} \mu(f \geq \lambda_k) \right\} \geq \rho^{-q^2(1-\theta_0)/s_0\theta_0} \left(\frac{C\|\nabla f\|_p}{\rho - 1} \right)^{\theta_0}.$$

As $\lambda^s \mu(f \geq \lambda) \leq \|f\|_s^s$, we get

$$\lambda_k^{-s} \rho^{qk} \|f\|_s^s \geq \rho^{-q^2(1-\theta_0)/s_0\theta_0} \left(\frac{C\|\nabla f\|_p}{\rho - 1} \right)^{\theta_0}.$$

Observe here that $\lambda_k = \lambda_k(\epsilon)$ depends on the small parameter ϵ. However, we can let ϵ tend to 0 in the above inequality. Choosing $k = k(f) - 1$ and observing that for this choice of k

$$\lambda_k(0) \geq \rho^k(\rho - 1) \geq \rho^{-2}(\rho - 1)\|f\|_\infty,$$

we obtain

$$\|f\|_\infty^{-s+q} \|f\|_s^s \geq (\rho - 1)^s \rho^{-2s+2q} \rho^{-q^2(1-\theta_0)/s_0\theta_0} \left(\frac{C\|\nabla f\|_p}{\rho - 1} \right)^{\theta_0}.$$

Rearranging, we obtain

$$\|f\|_\infty \leq \rho^2 (\rho - 1)^{-1} \left(\rho^{-q(1-\theta_0)/s_0\theta_0} C\|\nabla f\|_p \right)^{1/(1-s/q)} \|f\|_s^{1/(1-q/s)}.$$

Picking $\rho = 1 + 1/(1 + |q|)$ (recall that q is negative), we get

$$\|f\|_\infty \leq 4(1 + |q|) \left(e^{(1-\theta_0)/s_0\theta_0} C\|\nabla f\|_p \right)^{1/(1-s/q)} \|f\|_s^{1/(1-q/s)}.$$

This ends the proof of Theorem 3.2.7.

Corollary 3.2.8 *Assume that $(S^{*,\theta_0}_{r_0,s_0})$ is satisfied for some $0 < r_0, s_0 \le \infty$ and $0 < \theta_0 \le 1$ and let $q = q(r_0, s_0, \theta_0)$ be defined as in (3.2.1). Assume that $-\infty < q < 0$ (this forces $s_0 < r_0$). Then there exists a constant A such that*

$$\forall f \in \mathcal{C}^\infty_0(\Omega), \quad \|f\|_\infty \le A C \mu(\Omega)^{-1/q}\|\nabla f\|_p$$

for all bounded domains $\Omega \subset M$. Here C is the constant appearing in $(S^{*,\theta_0}_{r_0,s_0})$.

Indeed, for any finite $s > 0$, $\|f\|_s \le \|f\|_\infty \mu(\Omega)^{1/s}$ if f is supported in Ω. Hence, after simplifications, we have

$$\|f\|_\infty \le A C \mu(\Omega)^{-1/q}\|\nabla f\|_p.$$

3.2.6   Increasing p

In the last three sections, we studied the inequalities $(S^{*,\theta}_{r,s})$ and $(S^\theta_{r,s})$ for a fixed value of p. In order to discuss what happens when the parameter p is allowed to vary, let us introduce the notation

$$(S^{*,\theta}_{r,s}(p)) \quad \text{and} \quad (S^\theta_{r,s}(p))$$

to refer to these inequalities with $1 \le p < \infty$.

Theorem 3.2.9 *Assume that for some $1 \le p_0 < \infty$, $0 < r_0, s_0 \le \infty$, $0 < \theta_0 \le 1$, the inequality $(S^{*,\theta_0}_{r_0,s_0}(p_0))$ is satisfied. Assume also that $q_0 = q(r_0, s_0, \theta_0)$ defined at (3.2.1) satisfies $1/q_0 < 1/p_0$ (which is obviously satisfied if $q_0 < 0$ or $q_0 = \infty$). Let ν be defined by (3.2.2), that is, $1/q_0 = 1/p_0 - 1/\nu$. Then all the inequalities*

$$(S^\theta_{r,s}(p))$$

where $0 < r, s \le \infty$, $0 < \theta \le 1$, $p_0 \le p < \infty$, and

$$\frac{1}{q(r,s,\theta)} = \frac{1}{p} - \frac{1}{\nu}$$

are also satisfied.

By the results of the last three sections, we can assume without loss of generality that the inequality $(S^{\sigma_0}_{p_0,u_0}(p_0))$ is satisfied for some (or all) u_0, σ_0 such that $q(p_0, u_0, \sigma_0) = q_0$. For any $f \in \mathcal{C}^\infty_0(M)$, we can apply this inequality to $|f|^\gamma$, $\gamma \ge 1$,

$$\|f\|^\gamma_{p_0\gamma} \le C \left(\gamma \int_M |f|^{(\gamma-1)p_0}|\nabla f|^{p_0} d\mu \right)^{\sigma_0/p_0} \|f\|^{\gamma(1-\sigma_0)}_{u_0\gamma}.$$

By the Hölder inequality

$$\left(\int f^{(\gamma-1)r} g^r d\mu\right)^{1/r} \le \|f\|_{r\gamma}^{\gamma-1} \|g\|_{r\gamma}$$

this yields

$$\|f\|_{p_0\gamma}^{\gamma} \le C\gamma^{\sigma_0/p_0} \|f\|_{p_0}^{(\gamma-1)\sigma_0} \|\nabla f\|_{p_0\gamma}^{\sigma_0} \|f\|_{u_0\gamma}^{\gamma(1-\sigma_0)}.$$

Simplifying, we get that the inequality $(S_{p,u}^{\sigma}(p))$ is satisfied where $p = \gamma p_0$, with $\sigma = \sigma_0/(\sigma_0 + \gamma(1-\sigma_0))$, $u = p u_0/p_0$. Now, observe that

$$\frac{1-\sigma}{\sigma} = \gamma \frac{1-\sigma_0}{\sigma_0}.$$

It follows that the parameter ν defined by $1/\nu = 1/p_0 - 1/q_0$ satisfies

$$
\begin{aligned}
\frac{1}{\nu} &= \frac{1}{p_0} - \frac{1}{q_0} = \frac{1}{p_0} - \frac{1}{\sigma_0}\left(\frac{1}{p_0} - \frac{1-\sigma_0}{u_0}\right) \\
&= \frac{1-\sigma_0}{\sigma_0}\left(\frac{1}{u_0} - \frac{1}{p_0}\right) = \frac{1-\sigma}{\gamma\sigma}\left(\frac{1}{u_0} - \frac{1}{p_0}\right) \\
&= \frac{1-\sigma}{\sigma}\left(\frac{1}{\gamma u_0} - \frac{1}{\gamma p_0}\right) = \frac{1-\sigma}{\sigma}\left(\frac{1}{u} - \frac{1}{p}\right) \\
&= \frac{1}{p} - \frac{1}{\sigma}\left(\frac{1}{p} - \frac{1-\sigma}{u}\right) = \frac{1}{p} - \frac{1}{q(p,u,\sigma)}.
\end{aligned}
$$

In words, ν defined by $1/\nu = 1/p - 1/q$ has not changed when passing from $(S_{p_0,u_0}^{\sigma_0}(p_0))$ to $(S_{p,u}^{\sigma}(p))$.

In any case, we have that $(S_{p,u}^{\sigma}(p))$ is satisfied for some σ, u such that

$$\frac{1}{q(p,u,\sigma)} = \frac{1}{p} - \frac{1}{\nu}.$$

Now, depending on whether this q is positive, infinity, or negative, one of Theorems 3.2.2, 3.2.5, 3.2.7 shows that *all* the inequalities $(S_{r,s}^{\theta}(p))$, $0 < r, s \le \infty$, $0 < \theta \le 1$, with

$$\frac{1}{q(r,s,\theta)} = \frac{1}{p} - \frac{1}{\nu}$$

are satisfied. This is the desired conclusion.

Corollary 3.2.10 *Assume that for some* $1 \le p_0 < \infty$, $0 < r_0, s_0 \le \infty$, $0 < \theta_0 \le 1$, *the inequality* $(S_{r_0,s_0}^{*,\theta_0}(p_0))$ *is satisfied. Assume also that* $q_0 = q(r_0, s_0, \theta_0)$ *defined at* (3.2.1) *satisfies* $1/q_0 < 1/p_0$ *(which is obviously satisfied if* $q_0 < 0$ *or* $q_0 = \infty$*). Let* ν *be defined by* (3.2.2), *that is,* $1/q_0 = 1/p_0 - 1/\nu$. *Then there exists a constant* $c > 0$ *such that* $V(x,t) \ge ct^{\nu}$.

By Theorem 3.2.9, we can assume that $(S_{r,s}^\theta(p))$ is satisfied with p so large that $q(r,s,\theta) < 0$ (indeed, $1/q(r,s,\theta) = 1/p - 1/\nu$ with $0 < \nu < \infty$). By Theorem 3.2.7, we then have the inequality

$$\forall f \in \mathcal{C}_0^\infty(M), \quad \|f\|_\infty \le A\|\nabla f\|_p^\theta\|f\|_1^{1-\theta}$$

with $\theta = (1 - 1/p + 1/\nu)^{-1}$. Applying this to the function $f(y) = (t - d(x,y))_+$ yields

$$t \le AV(x,t)^{\theta/p}t^{1-\theta}V(x,t)^{1-\theta},$$

that is,

$$V(x,t)^{1/\nu} \ge A^{-1}t.$$

Corollary 3.2.10 is the same as Theorem 3.1.6. Note that the proof above does not use the fact that $\lim_{t\to 0} t^{-n}V(x,t) = \Omega_n > 0$ whereas the proof of Theorem 3.1.6 used this fact. Corollary 3.2.10 shows that the volume growth lower bound $V(x,t) \ge ct^\nu$ can be interpreted as a very weak form (a vestige) of Sobolev inequality.

3.2.7   Local versions

We would like to record here two useful comments about the results obtained in the previous four sections. The first comment is that we can replace M by an open subset, say $U \subset M$, without changing anything in Theorems 3.2.2, 3.2.5, 3.2.6, 3.2.7, 3.2.9, or in Lemma 3.2.3 or in Corollaries 3.2.4, 3.2.8. In other words, we do not need to assume that M is complete in these results. Note that Corollary 3.2.10 does not extend unchanged to the case of open subsets of M.

We would like also to mention that the results listed above can be extended to the case where the inequalities $(S_{r,s}^{*,\theta}(p))$, $(S_{r,s}^\theta(p))$ are replaced by their uniform local counterparts

$$\sup_{t>0}\left\{t\mu(\{|f| > t\})^{1/r}\right\} \le \left(C\|\nabla f\|_p + T\|f\|_p\right)^\theta \left(\|f\|_\infty\mu(\mathrm{supp}(f))^{1/s}\right)^{1-\theta},$$

$$\|f\|_r \le \left(C\|\nabla f\|_p + T\|f\|_p\right)^\theta\|f\|_s^{1-\theta}$$

for all $f \in \mathcal{C}_0^\infty(M)$, which we denote respectively by $(\widetilde{S}_{r,s}^{*,\theta}(p))$, $(\widetilde{S}_{r,s}^\theta(p))$.

This is more or less obvious once it has been observed that the quantity

$$W_p(f) = C\|\nabla f\|_p + T\|f\|_p$$

behaves just like $\|\nabla f\|_p$ with respect to the transformation $f \mapsto (f - t)_+ \wedge s = f_t^s$, $t,s > 0$. More precisely, we have $W_p(f_t^s) \le W(f)$ and

$$\sum_k W_p(f_{\rho,k}) \le W_p(f)$$

for $\rho > 1$ and $f_{\rho,k} = (f - \rho^k)_+ \wedge \rho^k(\rho - 1)$. Note that the ratio C/T is kept constant in this process.

Here are a few statements that are easily obtained by implementing this remark.

Theorem 3.2.11 *Let M be a complete Riemannian manifold. Assume that the inequality*

$$\sup_{t>0}\left\{t\mu(\{|f| > t\})^{1/r_0}\right\} \le (C\|\nabla f\|_p + T\|f\|_p)^{\theta_0}\left(\|f\|_\infty\mu(\mathrm{supp}(f))^{1/s_0}\right)^{1-\theta_0},$$

is satisfied for all $f \in \mathcal{C}_0^\infty(M)$ for some $0 < r_0, s_0 \le \infty$, $0 < \theta_0 \le 1$ and $1 \le p \le 1$. Define q by $1/r_0 = \theta_0/q + (1 - \theta_0)/s_0$ and assume that $p \le q < \infty$. Then there exists a constant A such that

$$\forall f \in \mathcal{C}_0^\infty(M), \quad \|f\|_q \le A\left(C\|\nabla f\|_p + T\|f\|_p\right).$$

Corollary 3.2.12 *For $\nu > 2$, the Nash-type inequality*

$$\forall f \in \mathcal{C}_0^\infty(M), \quad \|f\|_2^{(1+2/\nu)} \le (C\|\nabla f\|_2 + T\|f\|_2)\|f\|_1^{2/\nu}$$

implies the (L^2, ν)-Sobolev inequality

$$\forall f \in \mathcal{C}_0^\infty(M), \quad \|f\|_{2\nu/(\nu-2)} \le A_\nu\left(C\|\nabla f\|_2 + T\|f\|_2\right)$$

for some constant A_ν.

Theorem 3.2.13 *Assume that for some $1 \le p_0 < \infty$, $0 < r_0, s_0 \le \infty$, $0 < \theta_0 \le 1$, the inequality $(\widetilde{S}_{r_0,s_0}^{*,\theta_0}(p_0))$ is satisfied for some $C, T > 0$. Assume also that $q_0 = q(r_0, s_0, \theta_0)$ defined at (3.2.1) satisfies $1/q_0 < 1/p_0$ (which is obviously satisfied if $q_0 < 0$ or $q_0 = \infty$). Let ν be defined by (3.2.2), that is, $1/q_0 = 1/p_0 - 1/\nu$. Then there exists a constant $c > 0$ such that*

$$\forall x \in M, \ \forall t \in (0, T^{-1}), \quad V(x, t) \ge ct^\nu.$$

3.3 Examples

3.3.1 Pseudo-Poincaré inequalities

It turns out that a useful and widely applicable tool to prove Sobolev inequalities is a set of inequalities indexed by the semi-axis $\{t > 0\}$ that we call pseudo-Poincaré inequalities. For any $f \in \mathcal{C}_0^\infty(M)$, set

$$f_t(x) = \frac{1}{V(x,t)}\int_{B(x,t)} f(y)d\mu(y).$$

We say that M satisfies a pseudo-Poincaré inequality in L^p if there exists a constant C such that

$$\forall f \in \mathcal{C}_0^\infty(M), \ \forall t > 0, \quad \|f - f_t\|_p \leq C\,t\,\|\nabla f\|_p. \tag{3.3.1}$$

Note that the L^p norms on both side are taken over the whole space M. This should be interpreted as an inequality concerning the approximation of f by the more regular functions f_t, which are averages over balls of radius t. The important fact about f_t is that

$$\|f\|_\infty \leq V(x,t)^{-1}\|f\|_1. \tag{3.3.2}$$

This obvious fact should be thought of as a quantitative version of the "smoothing" effect of $f \to f_t$. Actually, in the arguments that will be developed below the only things that matter are that f_t satisfies both (3.3.1) and (3.3.2).

Theorem 3.3.1 *Assume that M satisfies (3.3.1) for some $1 \leq p_0 < \infty$ and that*

$$\inf_{\substack{t>0 \\ x \in M}} \left\{ t^{-\nu} V(x,t) \right\} > 0 \tag{3.3.3}$$

for some $\nu > 0$. Then the inequalities of Sobolev type $(S_{r,s}^\theta(p))$ hold true for all $p \geq p_0$ and all $0 < r, s \leq \infty$, $0 < \theta \leq 1$ such that $q(r,s,\theta)$ defined at (3.2.1) satisfies $1/q = 1/p - 1/\nu$.

The proof is surprisingly simple. It has its origins in [72]. By Theorem 3.2.9, it suffices to treat the case $p = p_0$. By hypothesis there exists a constant $c > 0$ such that $V(x,t) \geq (ct)^\nu$ for all $x \in M$ and $t > 0$. Fix $f \in \mathcal{C}_0^\infty(M)$, $f \geq 0$ and a real $\lambda > 0$. For any $t > 0$, write

$$\mu(f \geq \lambda) \leq \mu(|f - f_t| \geq \lambda/2) + \mu(f_t \geq \lambda/2)$$

and pick t so that $(ct)^{-\nu}\|f\|_1 = \lambda/4$. Then

$$\mu(f_t \geq \lambda/2) = 0$$

and

$$\begin{aligned}
\mu(f \geq \lambda) &\leq \mu(|f - f_t| \geq \lambda/2) \\
&\leq (2/\lambda)^p\|f - f_t\|_p^p \\
&\leq (2Ct\|\nabla f\|_p/\lambda)^p \\
&= \left[(2^{1+2/\nu}C/c)\|f\|_1^{1/\nu}\|\nabla f\|_p \lambda^{-1-1/\nu} \right]^p.
\end{aligned}$$

That is

$$\lambda^{p(1+1/\nu)}\mu(f \geq \lambda) \leq \left[(2^{1+2/\nu}C/c)\|f\|_1^{1/\nu}\|\nabla f\|_p \right]^p.$$

Raising this to the power τ/p with $\tau = 1/(1 + 1/\nu) = \nu/(1 + \nu)$ yields

$$\lambda\mu(f \geq \lambda)^{\tau/p} \leq 4(C/c)^{\tau}\|\nabla f\|_p^{\tau}\|f\|_1^{1-\tau}$$

which is $(S^{*\tau}_{p/\tau,1}(p))$. This, together with Theorems 3.2.2, 3.2.5, 3.2.7, proves Theorem 3.3.1.

We now state some obvious but important corollaries of Theorem 3.3.1.

Corollary 3.3.2 *Assume that M satisfies (3.3.1) with $p = 1$ and (3.3.3) for some $\nu > 1$. Then the isoperimetric inequality*

$$\mu(\Omega)^{1-1/\nu} \leq C_1\mu(\partial\Omega)$$

is satisfied by all bounded sets Ω with smooth boundary. Also the (L^1, ν)-Sobolev inequality

$$\forall f \in \mathcal{C}_0^{\infty}(M), \quad \|f\|_{\nu/(\nu-1)} \leq C_1\|\nabla f\|_1$$

is satisfied.

Corollary 3.3.3 *Fix $1 \leq p < \infty$ and $\nu > p$. Assume that M satisfies (3.3.1) and (3.3.3) for these p and ν. Then the (L^p, ν)-Sobolev inequality*

$$\forall f \in \mathcal{C}_0^{\infty}(M), \quad \|f\|_{p\nu/(\nu-p)} \leq C_2\|\nabla f\|_p$$

is satisfied.

Corollary 3.3.4 *Assume that M satisfies (3.3.1) with $p = 2$ and (3.3.3) for some $\nu > 0$. Then the Nash inequality*

$$\forall f \in \mathcal{C}_0^{\infty}(M), \quad \|f\|_2^{2(1+2/\nu)} \leq C_3\|\nabla f\|_2^2\|f\|_1^{4/\nu}$$

is satisfied.

3.3.2   Pseudo-Poincaré technique: local version

The argument developed in the section above can easily be adapted to the case when one only has local hypotheses. As an example, we prove the following result.

Theorem 3.3.5 *Fix $R > 0$. Assume that M satisfies*

$$\forall f \in \mathcal{C}_0^{\infty}(M), \ \forall t \in (0, R), \quad \|f - f_t\|_{p_0} \leq C\,t\,\|\nabla f\|_{p_0} \tag{3.3.4}$$

for some $1 \leq p_0 < \infty$ and that

$$\inf_{\substack{t \in (0,R) \\ x \in M}} \left\{ t^{-\nu}V(x, t) \right\} = c > 0 \tag{3.3.5}$$

for some $\nu > p_0$. Then the inequalities

$$\|f\|_q \leq [C(\nu, p)/c]\left(C\|\nabla f\|_p + R^{-1}\|f\|_p\right)$$

are satisfied for all $p_0 \leq p < \nu$ and $1/q = 1/p - 1/\nu$. The constant $C(\nu, p)$ is independent of C, c and R.

The proof is similar to that of Theorem 3.3.1. We can assume that $p = p_0$. By hypothesis, $V(x,t) \geq (ct)^\nu$ for all $x \in M$ and $t \in (0, R)$. Fix $f \in \mathcal{C}_0^\infty(M)$, $f \geq 0$ and a real $\lambda > 0$. For any $0 < t < R$, write

$$\mu(f \geq \lambda) \leq \mu(|f - f_t| \geq \lambda/2) + \mu(f_t \geq \lambda/2).$$

If possible, pick t so that $(ct)^{-\nu}\|f\|_1 = \lambda/4$. This can be done if $\lambda > 4(cR)^{-\nu}\|f\|_1$. Then, by (3.3.2)

$$\mu(f_t \geq \lambda/2) = 0$$

and

$$
\begin{aligned}
\mu(f \geq \lambda) &\leq \mu(|f - f_t| \geq \lambda/2) \\
&\leq (2/\lambda)^p \|f - f_t\|_p^p \\
&\leq (2Ct\|\nabla f\|_p/\lambda)^p \\
&= \left[(2^{1+2/\nu}C/c)\|f\|_1^{1/\nu}\|\nabla f\|_p \lambda^{-1-1/\nu}\right]^p.
\end{aligned}
$$

That is

$$\lambda^{p(1+1/\nu)}\mu(f \geq \lambda) \leq \left[(2^{1+2/\nu}C/c)\|f\|_1^{1/\nu}\|\nabla f\|_p\right]^p.$$

Now, if $\lambda \leq 4(cR)^{-\nu}\|f\|_1$, simply write

$$\mu(f > \lambda) \leq \lambda^{-p}\|f\|_p^p$$

to see that, in this case,

$$\lambda^{p(1+1/\nu)}\mu(f \geq \lambda) \leq \left[4^{1/\nu}(cR)^{-1}\|f\|_1^{1/\nu}\|f\|_p\right]^p.$$

Thus, in all cases,

$$\lambda^{p(1+1/\nu)}\mu(f \geq \lambda) \leq \left[(2^{1+2/\nu}/c)\|f\|_1^{1/\nu}(C\|\nabla f\|_p + R^{-1}\|f\|_p)\right]^p.$$

Raising this to the power τ/p with $\tau = 1/(1 + 1/\nu) = \nu/(1 + \nu)$ yields

$$\lambda\mu(f \geq \lambda)^{\tau/p} \leq 4(1/c)^\tau (C\|\nabla f\|_p + R^{-1}\|f\|_p)^\tau \|f\|_1^{1-\tau}$$

which is the local version of $(S_{p/\tau,1}^{*\tau}(p))$. This, together with the local version of Theorem 3.2.2, proves Theorem 3.3.5. See Section 3.2.7.

Since we have not explained in detail the local version of Theorem 3.2.2, let us note that in the case $p_0 = p = 1$, the argument above yields

$$\lambda\mu(f \geq \lambda)^\tau \leq C_1(C\|\nabla f\|_1 + R^{-1}\|f\|_1)^\tau \|f\|_1^{1-\tau}.$$

Applying this inequality to regularizations of $f = \mathbf{1}_\Omega$ (where Ω is a bounded set having smooth boundary) with $\lambda = 1/2$ yields the isoperimetric inequality

$$\mu(\Omega)^{1-1/\nu} \leq C_1'(C\mu(\partial\Omega) + R^{-1}\mu(\Omega)).$$

This inequality can then be integrated using the co-area formula to recover the Sobolev inequality

$$\|f\|_{\nu/(\nu-1)} \leq C_1'(C\|\nabla f\|_1 + R^{-1}\|f\|_1).$$

3.3.3 Lie groups

A connected Lie group of topological dimension n is a manifold G equipped with a product $G \times G \ni (g, h) \mapsto gh$ such that $(G, \cdot)$ is a group and the application $G \times G \ni (g, h) \mapsto gh^{-1} \in G$ is analytic. The fifth Hilbert problem was to decide if every connected locally Euclidean topological group is a Lie group. It was solved with an affirmative answer by Gleason, Montgomery and Zippin in 1952.

Any locally compact group carries some left (resp. right) invariant measures called left (resp. right) Haar measures. Any two left (resp. right) Haar measures μ_1, μ_2 are related by $\mu_1 = c\mu_2$ for some $c > 0$, so that essentially there is only one left (resp. right) Haar measure. Let μ be a left Haar measure on G. For $g \in G$ and A a Borel subset of G, let $\mu_g(A) = \mu(Ag)$. Then μ_g is another left Haar measure so that there exists $m(g) > 0$ such that $\mu_g = m(g)\mu$. Obviously, $m(gh) = m(g)m(h)$ and $m(\mathrm{id}) = 1$. The function m is called the modular function of G. It can be trivial (i.e., $m \equiv 1$), in which case we say that G is unimodular, or non-trivial. Obviously, G is unimodular if and only if any left Haar measure is also a right Haar measure.

Let T be the tangent space at the identity element id. By the left action of G on itself, any vector ξ in T defines a left-invariant vector field X_ξ on G. Conversely, any left-invariant vector field is determined by its value at id. That is, the space of all invariant vector fields on G (i.e., the Lie algebra of G) is isomorphic, as a vector space, to T. We can now fix a Euclidean metric on T and turn it into a left invariant Riemannian structure on G. If $(\xi_1, \ldots, \xi_n)$ is an orthonormal basis of T and $X_i = X_{\xi_i}$ then, at each point $g \in G$, $(X_1(g), \ldots, X_n(g))$ is an orthornormal basis of T_g for our Riemannian structure. By construction, the Riemannian measure associated to this structure must be left-invariant: it is a left Haar measure. Call it μ.

Given a function $f \in \mathcal{C}_0^\infty(M)$, ∇f is the vector field (not a left-invariant vector field!) defined by

$$df(Z) = \langle \nabla f, Z \rangle$$

for all vector fields Z. Computing in the orthonormal basis $(e_1, \ldots, e_n) = (X_1(g), \ldots, X_n(g))$ of T_g, $g \in G$, we have

$$df(Z)(g) = \sum_1^n z_i X_i f(g) = \sum_1^n z_i (\nabla f(g))_i$$

where $Z(g) = \sum z_i e_i$. That is, in coordinates,

$$\nabla f(g) = (X_1 f(g), \ldots, X_n f(g)). \tag{3.3.6}$$

Hence

$$|\nabla f(g)| = \sqrt{\sum_1^n |X_i f(g)|^2}.$$

We now want to compute the Laplace–Beltrami operator. Any left-invariant vector field X generates a one-parameter group $\{\phi_X(t) : t \in \mathbb{R}\}$ where $\phi_X : \mathbb{R} \to G$ satisfies $\phi'_X(t) = X(\phi_X(t))$. Moreover, for any function $f \in \mathcal{C}_0^\infty(M)$, Xf can be computed by

$$Xf(g) = \lim_{t \to 0} \frac{f(g\phi_X(t)) - f(g)}{t}.$$

Now, observe that if μ_r is a right Haar measure,

$$\int_G f_1(g\phi_X(t))f_2(g)d\mu_r(g) = \int_G f_1(g)f_2(g\phi_X^{-1}(t))d\mu_r(g).$$

As $\phi_X^{-1}(t) = \phi_X(-t)$, it follows that

$$\int_G Xf_1(g)f_2(g)d\mu_r(g) = - \int_G f_1(g)Xf_2(g)d\mu_r(g).$$

In particular this applies to the X_i's, which are thus skew symmetric with respect to μ_r. Let

$$Z(g) = \sum_1^n z_i(g)X_i(g)$$

be any smooth vector field with compact support on G. Then

$$\begin{aligned}
\int_G \langle Z, \nabla f \rangle d\mu_r &= \int_G \sum_1^n z_i(X_if)d\mu_r \\
&= - \int_G \left(\sum_1^n X_iz_i \right) f d\mu_r.
\end{aligned}$$

This shows that, *if G is unimodular*, i.e., $\mu_r = \mu$, then

$$\mathrm{div}(Z)(g) = \sum_1^n X_iz_i(g).$$

In this case, it follows that

$$\Delta f = - \mathrm{div}(\nabla f) = - \sum_1^n X_i^2 f. \tag{3.3.7}$$

If G is not unimodular, we can still compute the divergence by using the following trick. Let m be the modular function on G. Observe that $m^{-1}\mu$ is a right-invariant measure on G. Call it μ_r and write

$$\begin{aligned}
\int_G \langle Z, \nabla f \rangle d\mu &= \int_G \langle Z, \nabla f \rangle m d\mu_r = \int_G \sum_1^n z_i X_i f m d\mu_r \\
&= - \int_G \left[\sum_1^n X_i(mz_i) \right] f d\mu_r \\
&= - \int_G f \left(m \sum_1^n X_iz_i + \sum_1^n z_i X_i m \right) m^{-1} d\mu.
\end{aligned}$$

It is not hard to see that $X_i m = \lambda_i m$ for some constant λ_i. This is due to the fact that m is multiplicative. Indeed, for any left-invariant vector field X, we have

$$
\begin{aligned}
Xm(g) &= \lim_{t\to 0} \frac{m(g\phi_X(t)) - m(g)}{t} \\
&= \lim_{t\to 0} \frac{m(g)m(\phi_X(t)) - m(g)}{t} \\
&= m(g)\lim_{t\to 0} \frac{m(\phi_X(t)) - 1}{t} = \lambda_X m(g)
\end{aligned}
$$

with $\lambda_X = Xm(\mathrm{id})$. Thus, setting $\lambda_{X_i} = \lambda_i$, we get

$$
\int_G \langle Z, \nabla f\rangle d\mu = - \int_G f\left(\sum_1^n X_i z_i + \sum_1^n \lambda_i z_i\right) d\mu
$$

which gives

$$
\mathrm{div}(Z)(g) = \sum_1^n X_i z_i(g) + \sum_1^n \lambda_i z_i(g)
$$

and

$$
\Delta f = -\sum_1^n X_i^2 f - \sum_1^n \lambda_i X_i f. \tag{3.3.8}
$$

Let us now look at the Riemannian distance function. As the Riemannian structure we consider is left-invariant, the associated distance $(g,h) \mapsto d(g,h)$ is also left-invariant, i.e.,

$$
d(g,h) = d(\mathrm{id}, g^{-1}h).
$$

For all $g \in G$, we set

$$
|g| = d(\mathrm{id}, g)
$$

so that $d(g,h) = |g^{-1}h|$.

It follows from left invariance that all balls of a given radius have the same volume, that is,

$$
V(g,t) = V(\mathrm{id}, t) = V(t).
$$

3.3.4 Pseudo-Poincaré inequalities on Lie groups

This section shows that any unimodular Lie group satisfies the pseudo-Poincaré inequality (3.3.1). As a corollary, the results of Section 3.3.1 apply.

Theorem 3.3.6 *Let G be a unimodular Lie group equipped with a left-invariant Riemannian metric and the associated Haar measure μ. Then*

$$
\forall t > 0, \ \forall f \in \mathcal{C}_0^\infty(G), \quad \|f - f_t\|_p \le t\|\nabla f\|_p
$$

for all $1 \le p \le \infty$.

The proof is simple. Let $h \in G$. Let $\gamma_h : [0, t] \to G$ be a smooth curve in G such that $\gamma_h(0) = \text{id}$, $\gamma_h(t) = h$, and $|\dot{\gamma}_h(s)| \leq 1$. Note that $|h|$ is equal to the infimum of all $t > 0$ such that such a curve exists.

Now, for any $g \in G$ and $f \in \mathcal{C}_0^\infty(M)$,

$$f(gh) - f(g) = \int_0^t \partial_s f(g\gamma_h(s)) ds.$$

Thus

$$\begin{aligned} |f(gh) - f(g)| &\leq \int_0^t |df_{g\gamma_h(s)}(\partial_t g\gamma_h(s))| ds \\ &\leq \int_0^t |\nabla f(g\gamma_h(s))||\dot{\gamma}_h(s)| ds \leq \int_0^t |\nabla f(\gamma_h(s))| ds. \end{aligned}$$

This yields

$$|f(gh) - f(g)|^p \leq t^{p-1} \int_0^t |\nabla f(g\gamma_h(s))|^p ds$$

and

$$\begin{aligned} \int_G |f(gh) - f(g)|^p dg &\leq t^{p-1} \int_0^t \int_G |\nabla f(g\gamma_h(s))|^p dg\, ds \\ &= t^{p-1} \int_0^t \int_G |\nabla f(g)|^p dg\, ds \\ &= t^p \int_G |\nabla f(g)|^p dg. \end{aligned}$$

Note that here we have used the *right invariance* of the left Haar measure on G, i.e., the fact that G is unimodular. We can optimize over all curves joining the identity to h. This yields

$$\int_G |f(gh) - f(g)|^p dg \leq |h|^p \int_G |\nabla f(g)|^p dg.$$

Integrating this inequality in h over all $h \in B(t) = B(\text{id}, t)$ yields

$$\begin{aligned} \|f - f_t\|_p^p &= \int_G \left| f(g) - \frac{1}{V(t)} \int_{B(t)} f(gh) dh \right|^p dg \\ &\leq \frac{1}{V(t)} \int_B \int_G |f(g) - f(gh)|^p dg\, dh \\ &\leq \frac{1}{V(t)} \int_B |h| \|\nabla f\|_p^p dh \\ &\leq t^p \|\nabla f\|_p^p. \end{aligned}$$

This is the desired inequality.

Theorem 3.3.6 has the following corollary.

Corollary 3.3.7 *Let G be a unimodular Lie group equipped with a left-invariant Riemannian metric and the associated Haar measure μ. Assume that the volume growth function $V(t)$ satisfies*

$$\inf_{t>0} t^{-\nu} V(t) > 0$$

for some $\nu > 0$. Then G satisfies the isoperimetric inequality

$$\mu_n(\Omega)^{1-1/\nu} \leq C_1(\nu)\mu_{n-1}(\partial\Omega)$$

for all bounded sets $\Omega \subset G$ with smooth boundary.
For each $1 \leq p < \nu$, G sastisfies the (L^p, ν)-Sobolev inequality

$$\forall f \in \mathcal{C}_0^\infty(G), \quad \|f\|_{p\nu/(\nu-p)} \leq C(p,\nu)\|\nabla f\|_p.$$

More generally, for any $1 \leq p < \infty$ define $q(p)$ by $1/q(p) = 1/p - 1/\nu$. Then G satisfies all the inequalities $(S_{r,s}^\theta(p))$ for $1 \leq p < \infty$, $0 < s, r \leq \infty$ and $0 < \theta \leq 1$ such that

$$\frac{1}{r} = \frac{\theta}{q(p)} + \frac{1-\theta}{s}.$$

In particular, G sastisfies the Nash inequality

$$\forall f \in \mathcal{C}_0^\infty(G), \quad \|f\|_2^{2(1+2/\nu)} \leq C_2(\nu)\|\nabla f\|_2^2 \|f\|_1^{4/\nu}.$$

This corollary is a useful result because the volume growth of unimodular Lie groups is well understood. First, for $0 < t \leq 1$ we have

$$c_0 \leq t^{-n}V(t) \leq C_0$$

where n is the topological dimension of G. Second, by the work of Guivarc'h [36] (see also [45]), the volume growth function of any unimodular Lie group G satisfies the following alternative: either there exist $c, \alpha > 0$ such that for all $t \geq 1$,

$$V(t) \geq c \exp(\alpha t),$$

or there exist $0 < c \leq C < \infty$ and $d = 0, 1, 2, \ldots$ such that for all $t \geq 1$

$$c \leq t^{-d}V(t) \leq C.$$

A typical case where the volume of G has a polynomial behavior is when G is a simply connected nilpotent Lie group. In this case, the volume growth function V satisfies

$$\forall t > 1, \quad ct^d \leq V(t) \leq Ct^d$$

where d is an integer given by

$$d = \sum_1^k i \dim(\mathcal{G}_i/\mathcal{G}_{i+1}).$$

Here, the $\mathcal{G}_i$'s are subalgebras of the Lie algebra $\mathcal{G}$ of G defined inductively by $\mathcal{G}_1 = \mathcal{G}$, $\mathcal{G}_i = [\mathcal{G}, \mathcal{G}_{i-1}]$, $i = 2, \ldots$. The integer k is the smallest integer such that $\mathcal{G}_{k+1} = \{0\}$. That such a k exists is essentially the definition of G (i.e., $\mathcal{G}$) being nilpotent. As the topological dimension n of G is $n = \sum_1^k \dim(\mathcal{G}_i/\mathcal{G}_{i+1})$, it follows that any simply connected nilpotent Lie group G satisfies $n \leq d$. In particular, such a group has volume growth bounded below by

$$\forall t > 0, \quad V(t) \geq c_m t^m$$

for all $m \in [n, d]$.

For instance, the group of three by three upper-triangular matrices with diagonal entries equal to 1 (see (5.6.1)) is a simply connected nilpotent group known as the Heisenberg group and having $k = 2$ and $d = 4$. Of course, its topological dimension is $n = 3$. See [87] for details and references.

3.3.5   Ricci ≥ 0 and maximal volume growth

Let (M, g) be a complete Riemannian manifold. The Ricci curvature tensor $\mathcal{R}$ is a symmetric two-tensor obtained by contraction of the full curvature tensor. See, e.g., [13, 29]. Thus, it can be compared with the metric tensor g. Hypotheses of the type $\mathcal{R} \geq kg$, for some $k \in \mathbb{R}$, turn out to be sufficient to derive important analytic and geometric results. For instance, if $\mathcal{R} \geq kg$ with $k > 0$, then M must be compact. See, e.g., [13, Theorem 2.12]. If $k = 0$, the volume growth on (M, g) is at most Euclidean, that is, $\forall r > 0$, $V(x, r) \leq \Omega_n r^n$. See, e.g., [13, Theorem 3.9].

We want to show that complete manifolds of dimension n having non-negative Ricci curvature and maximal volume growth, that is, for which there exists $c > 0$ such that

$$\forall r > 0, \quad V(x, r) \geq cr^n,$$

satisfy the pseudo-Poincaré inequality (3.3.1).

Theorem 3.3.8 *Let (M, g) be a complete manifold of dimension n having non-negative Ricci curvature. Assume that there exists $c > 0$ such that*

$$\forall r > 0, \quad V(x, r) \geq cr^n.$$

Then

$$\forall r > 0, \ \forall f \in \mathcal{C}_0^\infty(M), \quad \|f - f_r\|_1 \leq (\Omega_n/c) \, r \, \|\nabla f\|_1.$$

Moreover, the Sobolev-type inequalities $(S_{r,s}^\theta(p))$ with

$$\frac{1}{r} = \theta \left(\frac{1}{p} - \frac{1}{n} \right) + \frac{1 - \theta}{s}, \quad 0 < r, s \leq \infty, \ 0 < \theta \leq 1$$

are all satisfied on M. In particular,

$$\forall f \in \mathcal{C}_0^\infty(M), \quad \|f\|_{np/(n-p)} \leq C_{n,p} \|\nabla f\|_p$$

for all $1 \leq p < n$.

Note that the hypothesis $V(x, r) \geq cr^n$ is also necessary for the last conclusion of this theroem to hold. By the results developed in this chapter, it suffices to prove the first assertion of the theorem. Observe that

$$\|f - f_r\|_1 = \int_M \left| f(x) - \frac{1}{V(x,r)} \int_{B(x,r)} f(y) dy \right| dx$$

$$\leq \int_M \int_M |f(x) - f(y)| \frac{\mathbf{1}_{B(x,r)}(y)}{V(x,r)} dy dx.$$

Now, consider the integral

$$\int_{B(x,r)} |f(x) - f(y)| dy.$$

To estimate this integral, we use polar (exponential) coordinates around x. See [13, Proposition 3.1]. This gives (in somewhat abusive notation)

$$\int_{B(x,r)} |f(x) - f(y)| dy = \int_0^r |f(x) - f(\rho, \theta)| \sqrt{\mathbf{g}}(\rho, \theta) d\rho d\theta$$

$$\leq \int_0^r \int_0^\rho |\partial_t f(t, \theta)| dt \sqrt{\mathbf{g}}(\rho, \theta) d\rho d\theta$$

$$\leq \int_0^r \int_0^\rho |\nabla f(t, \theta)| dt \sqrt{\mathbf{g}}(\rho, \theta) d\rho d\theta.$$

Here, we have simply used the usual trick to control $f(x) - f(y)$ by integrating along the geodesic segment from x to y and used polar exponential coordinates $y = (\rho, \theta)$ around x. In particular, $\sqrt{\mathbf{g}}(\rho, \theta) d\rho d\theta = dy$ is by definition the Riemannian volume element in polar coordinates. Strictly speaking, one should avoid the cut locus $C(x)$ of x in this computation. This however is not a problem because $C(x)$ has measure zero and $M \setminus C(x)$ is star shaped with respect to x. We refer the reader to [13] for a detailed treatment.

We now use the hypothesis that (M, g) has non-negative Ricci curvature. By Bishop's theorem [13, Theorem 3.8], the function $s \mapsto \sqrt{\mathbf{g}}(s, \theta)/s^{n-1}$ is non-increasing. It follows that

$$\int_{B(x,r)} |f(x) - f(y)| dy \leq \int_0^r \int_0^\rho |\nabla f(t, \theta)| t^{1-n} \sqrt{\mathbf{g}}(t, \theta) dt \rho^{n-1} d\rho d\theta$$

$$\leq \frac{r^n}{n} \int_0^r |\nabla f(t, \theta)| t^{1-n} \sqrt{\mathbf{g}}(t, \theta) dt d\theta$$

$$= \frac{r^n}{n} \int_{B(x,r)} \frac{|\nabla f(y)|}{d(x, y)^{n-1}} dy.$$

Using the hypothesis of maximal volume growth, we get

$$\int_M \int_M |f(x) - f(y)| \frac{\mathbf{1}_{B(x,r)}(y)}{V(x,r)} dy dx$$

$$\leq \frac{1}{cn} \int_M \int_{B(x,r)} \frac{|\nabla f(y)|}{d(x,y)^{n-1}} dy dx$$

$$\leq \frac{1}{cn} \int_M |\nabla f(y)| \int_{B(y,r)} \frac{1}{d(x,y)^{n-1}} dx dy.$$

Finally, by Bishop's theorem again, $\sqrt{\mathbf{g}}(t,\theta) \leq t^{n-1}$. It follows that

$$\int_{B(y,r)} \frac{1}{d(x,y)^{n-1}} dx \leq \omega_{n-1} r.$$

This yields

$$\|f - f_r\|_1 \leq \frac{\omega_{n-1}}{cn} r \|\nabla\|_1.$$

Recalling that $\Omega_n = \omega_{n-1}/n$, we see that this is the desired inequality. A variation on this proof yields a similar inequality in L^p norms.

The above proof of the pseudo-Poincaré inequality (3.3.1) for manifolds having non-negative Ricci curvature uses the additional hypothesis that the volume growth of M is maximal. One may wonder whether (3.3.1) holds without this additional hypothesis. The answer is yes.

Theorem 3.3.9 *Let (M,g) be a complete manifold of dimension n having non-negative Ricci curvature. Then*

$$\forall r > 0, \ \forall f \in \mathcal{C}_0^\infty(M), \ \|f - f_r\|_1 \leq 4^n r \|\nabla f\|_1.$$

To prove the desired inequality it suffices to bound

$$\int_M \int_M |f(x) - f(y)| \frac{\mathbf{1}_{B(x,r)}(y)}{V(x,r)} dy dx$$

by $C r \|\nabla f\|_1$. To make things more symmetric in x and y, note that

$$\frac{\mathbf{1}_{B(x,r)}(y)}{V(x,r)} \leq 2^n \frac{\mathbf{1}_{B(x,r)}(y)}{\sqrt{V(x,r)V(y,r)}}.$$

This uses the fact that $V(y,r) \leq V(x,2r) \leq 2^n V(x,r)$ if $d(x,y) \leq r$. The inequality $V(x,2r) \leq 2^n V(x,r)$ follows from the celebrated Bishop–Gromov comparison theorem (see, e.g., [13, 29]). Hence it suffices to bound

$$\int_M \int_M |f(x) - f(y)| \frac{\mathbf{1}_{B(x,r)}(y)}{\sqrt{V(y,r)V(x,r)}} dy dx.$$

Now, let $\gamma_{x,y}(t)$ be the geodesic from x to y parametrized by arc length (as in the proof of Theorem 3.3.8, we can ignore the cut locus). Then,

$$\int_M \int_M |f(x) - f(y)| \frac{\mathbf{1}_{B(x,r)}(y)}{\sqrt{V(y,r)V(x,r)}} dy dx$$

$$\leq \int_M \int_M \left[\int_0^{d(x,y)} |\nabla f(\gamma_{x,y}(t))| dt \right] \frac{\mathbf{1}_{B(x,r)}(y)}{\sqrt{V(y,r)V(x,r)}} dy dx.$$

By symmetry with respect to x, y, we can restrict integration in t to the interval $(d(x,y)/2, d(x,y))$ so that we obtain

$$\int_M \int_M |f(x) - f(y)| \frac{\mathbf{1}_{B(x,r)}(y)}{\sqrt{V(y,r)V(x,r)}} dy dx$$

$$\leq 2 \int_M \int_M \int_{d(x,y)/2}^{d(x,y)} \frac{|\nabla f(\gamma_{x,y}(t))| \mathbf{1}_{B(x,r)}(y)}{\sqrt{V(y,r)V(x,r)}} dt dy dx.$$

Now, by Bishop's theorem [13, Theorem 3.8], the Jacobian of the map $y \mapsto z = \gamma_{x,y}(t)$ is larger than 2^{-n+1} (see Lemma 5.6.7 for details) and

$$\frac{\mathbf{1}_{B(x,r)}(y)}{\sqrt{V(y,r)V(x,r)}} \leq 2^n \frac{\mathbf{1}_{B(x,r)}(\gamma_{x,y}(t))}{V(x,r)}.$$

Hence

$$\int_M \int_M |f(x) - f(y)| \frac{\mathbf{1}_{B(x,r)}(y)}{\sqrt{V(y,r)V(x,r)}} dy dx$$

$$\leq 2^{2n} \int_M \int_M \int_0^r \frac{|\nabla f(z)| \mathbf{1}_{B(x,r)}(z)}{V(x,r)} dt dz dx$$

$$\leq 2^{2n} r \int_M |\nabla f(z)| dz.$$

This yields the desired pseudo-Poincaré inequality. See also Theorem 5.6.6.

3.3.6   Sobolev inequality in precompact regions

Let (M, g) be a complete, non-compact, manifold of dimension n. It is often useful to invoke the fact that, if Ω is an open precompact subset of M, the usual $\mathbb{R}^n$ (local) Sobolev inequality $\|f\|_{pn/(n-p)} \leq C(\|\nabla f\|_p + \|f\|_p)$ holds for smooth functions with compact support in Ω, with a constant depending on Ω. Here we outline a direct proof of this fact.

Theorem 3.3.10 *Let (M, g) be a complete manifold of dimension n. For any open precompact region $\Omega \subset M$ and any $1 \leq p < n$, there exists a constant $C(\Omega, p)$ such that*

$$\forall f \in \mathcal{C}_0^\infty(\Omega), \quad \|f\|_{pn/(n-p)} \leq C(\Omega, p)(\|\nabla f\|_p + \|f\|_p).$$

We only need to prove the case $p = 1$. Now, the neigborhood $U = \{x \in M : d(x, \Omega) \leq 1\}$ of Ω is also precompact. Clearly, the volume function $V(x, t)$ satisfies

$$\inf_{\substack{x \in U \\ t \in (0,1)}} t^{-n} V(x, t) > 0.$$

Moreover, the volume element in polar coordinates around x, which we denote by $\sqrt{\mathbf{g}}_x(t,\theta)$ as in the previous section, satisfies

$$\forall\, x \in U, \quad \forall t \in (0,1), \quad c \leq t^{1-n}\sqrt{\mathbf{g}}_x(t,\theta) \leq C.$$

(This must be understood outside the cut locus of x. In fact, one can avoid the cut locus completely in this argument by considering only $t \in (0, \epsilon)$ with ϵ small enough). See [13]. With these observations we can use the proof technique of Section 3.3.5 to show that, for all $0 < t < 1$ and all $f \in \mathcal{C}_0^\infty(\Omega)$,

$$\|f - f_t\|_1 \leq C(U)\, t\, \|\nabla f\|_1.$$

Applying Theorem 3.3.5 yields the desired result. In general, one cannot dispense with the $\|f\|_p$ term appearing on the right-hand side in Theorem 3.3.10 (to see this consider constant functions on a compact manifold). However, if Ω is a relatively compact set in a complete *non-compact* manifold then the technique of Section 3.3.5 can be used to show that, for all $f \in \mathcal{C}_0^\infty(\Omega)$,

$$\|f\|_p \leq C'(\Omega,p)\|\nabla f\|_p$$

(a Poincaré inequality). Thus, one has the following result.

Theorem 3.3.11 *Let (M,g) be a complete non-compact manifold of dimension n. For any open precompact region $\Omega \subset M$ and any $1 \leq p < n$, there exists a constant $C(\Omega,p)$ such that*

$$\forall\, f \in \mathcal{C}_0^\infty(\Omega), \quad \|f\|_{pn/(n-p)} \leq C(\Omega,p)\|\nabla f\|_p.$$

Chapter 4

Two applications

4.1 Ultracontractivity

This section presents a few basic applications of Sobolev inequalities to the study of the heat diffusion semigroup $H_t = e^{-t\Delta}$. We show how Nash inequality implies that H_t acts as a bounded operator from L^1 to L^∞ with an explicit time dependent estimate. Basic references for this material are [67] and [11, 21, 87].

4.1.1 Nash inequality implies ultracontractivity

For the purpose of this section, let M be a locally compact, second countable, Hausdorff space equipped with a Borel measure μ. Let Q be a Dirichlet form on $L^2(M, d\mu)$ with domain $\mathcal{D}$. We refer to [21] for a short introduction and to [26] for a detailed treatment. The reader who does not want to learn about Dirichlet forms can think of the following basic example.

EXAMPLE Let M be a complete Riemannian manifold with Riemannian measure dx. Let $W^{1,2}(M) \subset L^2(M, dx)$ be the completion of $\mathcal{C}_0^\infty(M) \subset L^2(M, dx)$ for the norm $\sqrt{\|f\|_2^2 + \|\nabla f\|_2^2}$. Let $V > 0$ be a smooth bounded function on M and set $d\mu(x) = V(x)dx$. Then

$$Q(f, f) = \int_M |\nabla f(x)|^2 dx, \quad f \in W^{1,2}(M) \cap L^2(M, d\mu)$$

is a Dirichlet form on $L^2(M, d\mu)$.

Returning to the general case, any Dirichlet form is associated canonically to a self-adjoint semigroup of operators $H_t = e^{-tA}$, $t > 0$, on $L^2(M, \mu)$ generated by a possibly unbounded self-adjoint operator $-A$ with domain $\mathcal{D}(A) \subset \mathcal{D}$. In terms of the semigroup $(H_t)_{t>0}$, the domain of A is the space $\mathcal{D}(A)$ of all $f \in L^2(M, \mu)$ such that

$$\lim_{t \to 0} \frac{1}{t}(H_t f - f)$$

exists in $L^2(M, \mu)$, and A is given by

$$-Af = \lim_{t \to 0} \frac{1}{t}(H_t f - f).$$

Very generally, one can show by elementary arguments that if $f \in \mathcal{D}(A)$ then the function $x \mapsto H_t f(x)$ is in $\mathcal{D}(A)$ and $(t, x) \mapsto u(t, x) = H_t f(x)$ satisfies

$$(\partial_t + A)u = 0, \quad u(0, \cdot) = f. \tag{4.1.1}$$

Moreover, $\mathcal{D}(A)$ is dense in $L^2(M, d\mu)$. See, e.g., [69]. In the case of a self-adjoint semigroup on $L^2(M, d\mu)$ as above, one can appeal to spectral theory to show that, for any $f \in L^2(M, d\mu)$, $H_t f$ is in $\mathcal{D}(A)$ and $u(t, x) = H_t f(x)$ satisfies (4.1.1).

In terms of the form $(Q, \mathcal{D})$, the generator $-A$ and its domain can be defined by

$$\forall f \in \mathcal{D}(A), \ \forall g \in \mathcal{D}, \quad Q(f, g) = \int_M (Af)\overline{g}d\mu.$$

Because Q is a Dirichlet form, H_t takes real functions to real functions and so does A. Hence in most instance it suffices to work with real-valued functions. In the sequel, we always assume that all functions are real-valued unless it is explicitly stated that functions might be complex-valued. On at least one occasion we will indeed need to consider complex-valued functions.

In fact, not only does H_t preserve real functions but it also preserves positivity. It follows that H_t extends as a (weakly continuous) semigroup of contractions on $L^\infty(M, \mu)$. It also extends as a strongly continuous semigroup of contractions on all L^p spaces, $1 \leq p < \infty$.

A simple important fact is that

$$t \mapsto -\langle \partial_t H_t f, f \rangle = \|A^{1/2} H_{t/2} f\|_2^2 = \|H_{t/2} A^{1/2} f\|_2^2$$

is non-increasing for all $f \in \mathcal{D}$. This implies that

$$\|f\|_2^2 - \langle H_t f, f \rangle \geq -t\langle \partial_t H_t f, f \rangle.$$

Hence $\|A^{1/2} H_t f\|_2 \leq (2t)^{-1/2}\|f\|_2$ for all $f \in \mathcal{D}$ and thus for all $f \in L^2(M, \mu)$. By the semigroup property, it follows that $H_t f$ is in the domain of $A^{k/2}$ for all integers k and all $f \in L^2(M, d\mu)$ and

$$\|A^{k/2} H_t\|_{2 \to 2} \leq (2t/k)^{-k/2}. \tag{4.1.2}$$

EXAMPLE   In the case of our basic example (i.e., when M is a Riemannian manifold, $Q(f, f) = \int |\nabla f|^2 dx$ and $d\mu(x) = V(x)dx$, V is positive and smooth, and dx is the Riemannian measure), we have $A = V^{-1}\Delta$ and the semigroup H_t leads to solutions of the equation

$$(\partial_t + V^{-1}\Delta)u = 0.$$

Observe that positivity preserving and the fact that H_t contracts L^∞ can be seen as consequences of the (parabolic) maximum principle satisfied by such equations. The local theory of parabolic equations associated to elliptic operators implies that any solution u of $(\partial_t + V^{-1}\Delta)u = 0$ is smooth. In particular, for any $f \in L^1(M, \mu)$, $H_t f \in \mathcal{C}^\infty(M)$. It follows that

$$H_t f(x) = \int_M h(t, x, y) f(y) d\mu(y)$$

where $h(t, x, y)$ is a non-negative smooth function which we call the heat kernel associated to $V^{-1}\Delta$ on $L^2(M, d\mu)$. When $V \equiv 1$, this is the usual Riemannian heat diffusion kernel (heat kernel for short) of M. When $M = \mathbb{R}^n$ and $V \equiv 1$ then

$$h(t, x, y) = \frac{1}{(4\pi t)^{n/2}} \exp(-|x - y|^2/4t).$$

Returning to the general case, as H_t preserves positivity and contracts $L^\infty(M, \mu)$, it follows that for each t, x there exists a measure $h(t, x, dy)$ (often called the transition function of H_t) such that

$$H_t f(x) = \int_M f(y) h(t, x, dy)$$

and

$$h(t, x, M) \leq 1.$$

This observation can be used to show that H_t contracts $L^p(M, \mu)$, that is,

$$\|H_t\|_{p \to p} \leq 1$$

for all $1 \leq p \leq \infty$. Indeed, by Jensen's inequality,

$$|H_t f(x)|^p \leq [H_t(|f|^p)](x).$$

Thus, it is enough to show that H_t contracts $L^1(M, \mu)$. This follows by duality from the fact that

$$\int_M g H_t f d\mu = \int_M f H_t g d\mu \leq \|g\|_\infty \|f\|_1$$

for all $f, g \in L^1(M, \mu) \cap L^\infty(M, \mu)$.

There is no reason, in general, that H_t should send $L^1(M, \mu)$ into any L^p space, $p > 1$. However, in many important cases, $(H_t)_{t>0}$ has the qualitative smoothing effect that $H_t f$ is bounded for any $f \in L^1(M, d\mu)$ and any $t > 0$. When this is the case, for any $t > 0$, there exists a constant C_t such that

$$\forall f \in L^1(M, \mu), \quad \|H_t f\|_\infty \leq C_t \|f\|_1$$

and one says that H_t is ultracontractive.

Obviously, this is equivalent to saying that the measure $h(t, x, dy)$ is absolutely continuous with respect to μ and has a bounded density. In this case, we will write

$$h(t, x, dy) = h(t, x, y)d\mu(y).$$

With this notation, the properties

$$\forall t > 0, \quad \forall f \in L^1(M, \mu), \quad \|H_t f\|_\infty \leq C_t \|f\|_1$$

and

$$\forall t > 0, \quad \sup_{x,y \in M} h(t, x, y) \leq C_t$$

are equivalent.

The following elegant result was first proved by Nash [67] in the case where A is a symmetric uniformly elliptic second order differential operator in divergence form in $\mathbb{R}^n$.

Theorem 4.1.1 *Let Q be a Dirichlet form on $L^2(M, \mu)$ with associated semigroup $(H_t)_{t>0}$. Assume that the Nash inequality*

$$\forall f \in \mathcal{D}, \quad \|f\|_2^{2(1+2/\nu)} \leq CQ(f, f)\|f\|_1^{4/\nu}$$

is satisfied for some $\nu > 0$. Then $(H_t)_{t>0}$ is ultracontractive and

$$\forall t > 0, \quad \|H_t\|_{1 \to \infty} \leq (C\nu/2t)^{\nu/2},$$

or, equivalently, $(H_t)_{t>0}$ admits a density w.r.t. μ which satisfies

$$\forall t > 0, \quad \sup_{x,y \in M} \{h(t, x, y)\} \leq (C\nu/2t)^{\nu/2}.$$

Let $f \in L^1(M, \mu) \cap L^2(M, \mu)$, $\|f\|_1 = 1$. Then $\|H_t f\|_1 \leq 1$ for all $t > 0$. Recall that $H_t f \in \mathcal{D}(A)$ for all $t > 0$ where $-A$ is the infinitesimal generator. Set

$$u(t) = \|H_t f\|_2^2.$$

Then u has derivative

$$u'(t) = 2\langle \partial_t H_t f, H_t f \rangle$$

where the scalar product is in $L^2(M, \mu)$. It follows that

$$u'(t) = -2\langle AH_t f, H_t f \rangle = -2Q(H_t f, H_t f).$$

Thus, by the postulated Nash inequality and the fact that $\|H_t f\|_1 \leq 1$,

$$u(t)^{1+2/\nu} \leq -(C/2)u'(t).$$

Setting $v(t) = (\nu/2)u(t)^{-2/\nu}$, this yields $v'(t) \geq 2/C$. Hence

$$v(t) \geq (2/C)t,$$

that is,

$$u(t) \leq (C\nu/4t)^{\nu/2}.$$

This means that

$$\|H_t f\|_2 \leq (C\nu/4t)^{\nu/4}\|f\|_1$$

for all $f \in L^1(M, \mu)$ because $L^1(M, \mu) \cap L^2(M, \mu)$ is dense in $L^1(M, \mu)$. In other words

$$\|H_t\|_{1 \to 2} \leq (C\nu/4t)^{\nu/4}.$$

As H_t is self-adjoint, it follows by duality that

$$\|H_t\|_{2 \to \infty} \leq (C\nu/4t)^{\nu/4}.$$

By the semigroup property,

$$\|H_t\|_{1 \to \infty} \leq \|H_{t/2}\|_{1 \to 2}\|H_{t/2}\|_{2 \to \infty}$$

and we obtain

$$\|H_t\|_{1 \to \infty} \leq (C\nu/2t)^{\nu/2}$$

as desired.

For later applications, let us observe that the ultracontractivity of the semigroup $(H_t)_{t>0}$ implies that the time derivatives of the kernel $h(t, x, y)$ are also well behaved. Namely, by (4.1.2), under the hypotheses of Theorem 4.1.1, we also have

$$\|\partial_t^k H_t\|_{1 \to \infty} \leq C_k t^{-\nu/2-k} \tag{4.1.3}$$

and thus

$$\sup_{x,y \in M} \{|\partial_t^k h(t, x, y)|\} \leq C_k t^{-\nu/2-k}. \tag{4.1.4}$$

4.1.2 The converse

Carlen *et al.* [11] observed that Nash's result admits the following converse.

Theorem 4.1.2 *Let Q be a Dirichlet form with associated Markov semigroup $(H_t)_{t>0}$. Assume that there exists $\nu > 0$ such that*

$$\forall t > 0, \quad \|H_t\|_{1 \to 2} \leq (C/t)^{\nu/4}.$$

Then the Nash inequality

$$\forall f \in \mathcal{D}, \quad \|f\|_2^{2(1+2/\nu)} \leq C_\nu C Q(f, f)\|f\|_1^{4/\nu}$$

is satisfied with

$$C_\nu = 2(1 + 2/\nu)(1 + \nu/2)^{2/\nu}.$$

Let $f \in \mathcal{D} \cap L^1(M, \mu)$. Then we have

$$\|f\|_2^2 = \|H_t f\|_2^2 - \int_0^t \partial_s \|H_s f\|_2^2 ds.$$

Hence

$$\|f\|_2^2 = \|H_t f\|_2^2 + 2 \int_0^t \langle A H_s f, H_s f \rangle ds.$$

Now, we observe that $s \mapsto \langle H_s f, A H_s f \rangle$ is a non-increasing function of s because

$$\langle H_s f, A H_s f \rangle = \|A^{1/2} H_s f\|_2^2 = \|H_s A^{1/2} f\|_2^2.$$

Hence

$$\|f\|_2^2 \leq \|H_t f\|_2^2 + 2t \langle f, A f \rangle \leq (C/t)^{\nu/2} \|f\|_1^2 + 2t Q(f, f).$$

Optimize in $t > 0$ by choosing t such that

$$\nu C^{\nu/2} \|f\|_2 \|f\|_1^2 = 4 Q(f, f) t^{1 + \nu/2}.$$

After some algebra this yields

$$\|f\|_2^{2(1 + 2/\nu)} \leq 2(1 + 2/\nu)(1 + \nu/2)^{2/\nu} C Q(f, f) \|f\|_1^{4/\nu}.$$

The result of Carlen *et al.* and Nash's semigroup estimate prove the equivalence between Nash inequality and a specific heat kernel bound as stated in the following theorem.

Theorem 4.1.3 *Let Q be a Dirichlet form on $L^2(M, \mu)$ and let $(H_t)_{t>0}$ be the corresponding Markov semigroup. The following two properties are equivalent.*

(i) *The form Q satisfies the Nash inequality*

$$\forall f \in \mathcal{D}, \quad \|f\|_2^{2(1 + 2/\nu)} \leq C_1 Q(f, f) \|f\|_1^{4/\nu}.$$

(ii) *The semigroup $(H_t)_{t>0}$ is ultracontractive and its kernel $h(t, x, y)$ satisfies*

$$\forall t > 0, \quad \sup_{x,y} \{ h(t, x, y) \} \leq C_2 t^{-\nu/2}.$$

This equivalence is a very useful tool. It shows a certain stability of the decay condition $\sup_{x,y} \{ h(t, x, y) \} \leq C t^{-\nu/2}$ under perturbations of the generator that preserve the size of $Q(f, f)$ and the size of the measure μ. As a very simple example consider a complete manifold M and two measures $d\mu = V(x)dx$ and $d\mu'(x) = V'(x)dx$ with V, V' positive and smooth. Let $Q(f, f) = \int |\nabla f(x)|^2 dx$ and consider Q as a Dirichlet form on either $L^2(M, d\mu)$ or $L^2(M, d\mu')$. This yields two semigroups and two heat kernels h and h'. Assuming that $0 < c \leq V/V' \leq C < \infty$, if $\sup_{x,y} \{ h(t, x, y) \} \leq C t^{-\nu/2}$, the same must be true for $h'(t, x, y)$. Quasi-isometric changes of metric can be treated similarly. Proving such stability without the help of a result such as Theorem 4.1.3 is not easy.

4.2 Gaussian heat kernel estimates

This section presents in the Riemannian setting some Gaussian estimates which complement the ultracontractivity bounds of Section 4.1.1. In the presentation given below, Nash-type inequalities enter the argument only through the use of the ultracontractivity bound of Theorem 4.1.1. This approach is adapted from [41].

4.2.1 The Gaffney–Davies L^2 estimate

Let M be a complete non-compact Riemannian manifold equipped with its Riemannian measure μ. Let Δ be the Laplacian on M and let $h(t, x, y)$ be the heat diffusion kernel on M. That is, for each $x \in M$, $h(t, x, y) = u(t, y)$ is the minimal solution of

$$\begin{cases} (\partial_t + \Delta)u = 0 \\ u(0, y) = \delta_x(y). \end{cases}$$

Equivalently, $h(t, x, y)$ is the kernel (with respect to μ) of the semigroup $H_t = e^{-t\Delta}$ associated to the Dirichlet form $Q(f, f) = \int |\nabla f|^2 d\mu$ with domain $W_2^1(M)$, the closure of $\mathcal{C}_0^\infty(M)$ for the norm $\sqrt{\|f\|_2^2 + \|\nabla f\|_2^2}$.

For any function $\phi \in \mathcal{C}_0^\infty(M)$ with $\|\nabla \phi\|_\infty \leq 1$ and any complex number α, consider the semigroup defined by

$$H_t^{\alpha,\phi} f(x) = e^{-\alpha\phi(x)} \int h(t, x, y) e^{\alpha\phi(y)} f(y) dy = e^{-\alpha\phi(x)} H_t(e^{\alpha\phi} f)(x).$$

It is clear that this is a well-defined semigroup of operators on the spaces $L^p(M, \mu)$. Its infinitesimal generator is given by

$$-A_{\alpha,\phi} f = -e^{-\alpha\phi} \Delta(e^{\alpha\phi} f).$$

When α is real this semigroup preserves positivity but there is no reason that it contracts $L^p(M, \mu)$, for any $1 \leq p \leq \infty$. It is not self-adjoint but its adjoint is simply $H_t^{-\alpha,\phi}$.

The next lemma estimates the norm of these semigroups on $L^2(M, \mu)$. It is due to Gaffney [27] and was later rediscovered by Davies, who turned it into a powerful tool to derive pointwise Gaussian estimates under ultra-contractivity hypotheses. See [21] and the references therein.

Lemma 4.2.1 *For any function $\phi \in \mathcal{C}_0^\infty(M)$ with $\|\nabla \phi\|_\infty \leq 1$ and any real number α, the semigroup $(H_t^{\alpha,\phi})_{t>0}$ satisfies*

$$\forall t > 0, \quad \|H_t^{\alpha,\phi}\|_{2\to 2} \leq e^{\alpha^2 t}.$$

Set $u(t) = \|H_t^{\alpha,\phi} f\|_2^2$, $f \in L^2(M, \mu)$. Then u has derivative

$$u'(t) = -2\langle A_{\alpha,\phi} H_t^{\alpha,\phi} f, H_t^{\alpha,\phi} f \rangle.$$

Thus, it suffices to show that

$$\langle A_{\alpha,\phi}f, f \rangle \geq -\alpha^2 \|f\|_2^2 \tag{4.2.1}$$

for all $f \in \mathcal{C}_0^\infty(M)$. For, if this holds, we have $u' \leq \alpha^2 u$, that is, $u(t) \leq e^{\alpha^2 t}u(0)$ and the desired conclusion easily follows.

To prove (4.2.1), write

$$
\begin{aligned}
\langle A_{\alpha,\phi}f, f \rangle &= \langle e^{-\alpha\phi}\Delta(e^{\alpha\phi}f), f \rangle \\
&= \int \nabla(e^{\alpha\phi}f) . \nabla(e^{-\alpha\phi}f)d\mu \\
&= \int |\nabla f|^2 d\mu - \alpha^2 \int |\nabla\phi|^2|f|^2 d\mu \\
&\geq -\alpha^2 \|f\|_2^2.
\end{aligned}
$$

The last step above uses the hypothesis $|\nabla\phi| \leq 1$. This proves Lemma 4.2.1.

The following version of Gaffney's lemma dealing with time derivatives is also useful.

Lemma 4.2.2 *For all functions* $\phi \in \mathcal{C}_0^\infty(M)$ *with* $\|\nabla\phi\|_\infty \leq 1$ *and all* $\alpha \in \mathbb{R}$, *for all* $t > 0$ *and all* $\zeta = e^{i\tau} \in \mathbb{C}$, $\tau \in \mathbb{R}$, *with* $|\tau| \leq \epsilon$, $0 < \epsilon \leq \pi/4$, *we have*

$$\|H_{t\zeta}^{\alpha,\phi}\|_{2\to2} \leq e^{\alpha^2(1+\epsilon)t}.$$

In particular, there exists a constant C *such that*

$$\|\partial_t^k H_t^{\alpha,\phi}\|_{2\to2} \leq C^k k!(\epsilon t)^{-k} e^{\alpha^2(1+\epsilon)t}. \tag{4.2.2}$$

For the proof, we need to consider complex-valued functions (because the complex time semigroup does not preserve real functions). We denote by $\Re(z)$ the real part of any complex number z. For two complex-valued functions f, g, we set $\langle f, g \rangle = \int f \bar{g} \, d\mu$. Let $\zeta = \cos\tau + i\sin\tau \in \mathbb{C}$ with $|\tau| \leq \pi/4$. For any complex-valued $f \in \mathcal{C}_0^\infty(M)$, set

$$u(t) = \|H_{t\zeta}^{\alpha,\phi}f\|_2^2.$$

Then

$$u'(t) = -2\Re\left(\zeta\langle A_{\alpha,\phi}H_{t\zeta}^{\alpha,\phi}f, H_{t\zeta}^{\alpha,\phi}f \rangle\right).$$

Moreover,

$$
\begin{aligned}
\langle A_{\alpha,\phi}f, f \rangle &= \langle e^{-\alpha\phi}\Delta(e^{\alpha\phi}f), f \rangle \\
&= \int \nabla(e^{\alpha\phi}f) . \nabla(e^{-\alpha\phi}\bar{f})d\mu \\
&= \int |\nabla f|^2 d\mu - \alpha^2 \int |\nabla\phi|^2|f|^2 d\mu \\
&\quad + \alpha\left(\int f\nabla\phi . \nabla\bar{f}d\mu - \int \bar{f}\nabla\phi . \nabla f d\mu\right).
\end{aligned}
$$

The last term of this sum is purely imaginary and has modulus bounded by

$$\alpha^2 \int |\nabla \phi|^2 |f|^2 d\mu + \int |\nabla f|^2 d\mu.$$

Using $|\nabla \phi| \leq 1$, we obtain

$$-2\Re\left(\zeta \langle A_{\alpha,\phi} f, f \rangle\right) \leq -2(\cos\tau - |\sin\tau|) \int |\nabla f|^2 d\mu$$

$$+2(\cos\tau + |\sin\tau|)\alpha^2 \int |f|^2 d\mu$$

$$\leq 2(1 + |\tau|)\alpha^2 \int |f|^2 d\mu.$$

From this it follows that $u' \leq 2(1+\epsilon)\alpha^2 u$ and the first conclusion in Lemma 4.2.2 follows. The bound (4.2.2) is then obtained by applying Cauchy's formula to the holomorphic function $z \to H_z^{\alpha,\phi} f$ on a circle of radius $\epsilon t/10$ centered on the real axis at $t > 0$.

4.2.2 Complex interpolation

In this section we implicitly work with complex L^p-spaces in order to use complex interpolation techniques. Let us start with the following simple lemma.

Lemma 4.2.3 *Assume that the heat diffusion semigroup* $(H_t)_{t>0}$ *satisfies*

$$\forall t > 0, \quad \|H_t\|_{2\to\infty} \leq (C/t)^{\nu/4}.$$

Then, for all $2 \leq p \leq \infty$*, we also have*

$$\forall t > 0, \quad \|H_t\|_{p\to\infty} \leq (C/t)^{\nu/2p}.$$

By classical interpolation theory (see, e.g., [79, 81]), the bounds

$$\|H_t\|_{\infty\to\infty} \leq 1$$

and

$$\|H_t\|_{2\to\infty} \leq (C/t)^{\nu/4}$$

imply

$$\|H_t\|_{p\to\infty} \leq (C/t)^{\theta\nu/4}$$

if $1/p = \theta/2 + (1-\theta)/\infty$, i.e., $\theta = 2/p$. This is the desired result.

We will now need a more sophisticated version of the classical Riesz–Thorin interpolation theorem. In what follows, $i^2 = -1$, complex numbers are written $z = a + bi$, $a, b \in \mathbb{R}$ and we set $\Re(z) = a$. Let T_z be an analytic family of linear operators defined on $\mathcal{C}_0^\infty(M)$ for all complex z with

$0 \leq \Re(z) \leq 1$ (the analyticity hypothesis made here is that $z \mapsto \int g T_z f d\mu$ is analytic in the strip $0 < \Re(z) < 1$ and continuous in $0 \leq \Re(z) \leq 1$, for all complex-valued functions $f, g \in \mathcal{C}_0^\infty(M)$). Assume that $\|T_{bi}\|_{p_1 \to q_1} \leq M_1$ and $\|T_{1+bi}\|_{p_2 \to q_2} \leq M_2$ for all reals b and some fixed $1 \leq p_1, p_2, q_1, q_2 \leq \infty$. Then the interpolation theorem of E. Stein (which generalizes the classical Riesz-Thorin interpolation theorem to *analytic families* of operators) gives the bound

$$\forall \theta \in [0,1], \quad \|T_\theta\|_{p_\theta \to q_\theta} \leq M_1^{1-\theta} M_2^\theta$$

where

$$\frac{1}{p_\theta} = \frac{1-\theta}{p_1} + \frac{\theta}{p_2}, \quad \frac{1}{q_\theta} = \frac{1-\theta}{q_1} + \frac{\theta}{q_2}.$$

See [80, 81, 82].

Now, we consider the family of operators $H_t^{\alpha z, \phi}$ where α, ϕ, t are fixed, α is real and z is complex with $\Re(z) \in [0,1]$. Because multiplication by a complex function of modulus 1 is a contraction operator on $L^p(M, \mu)$, it is easy to see that

$$\|H_t^{\alpha z, \phi}\|_{2\to2} \leq \|H_t^{\alpha \Re(z), \phi}\|_{2\to2}.$$

In particular, by Lemma 4.2.1,

$$\|H_t^{\alpha(1+bi), \phi}\|_{2\to2} \leq e^{\alpha^2 t}.$$

Also,

$$\|H_t^{\alpha bi, \phi}\|_{\infty\to\infty} \leq 1.$$

By Stein's interpolation theorem [81], we conclude that

$$\|H_t^{2\alpha/p, \phi}\|_{p\to p} \leq e^{2\alpha^2 t/p}$$

for all $p \geq 2$. Changing α to $\alpha p/2$ yields

$$\|H_t^{\alpha, \phi}\|_{p\to p} \leq e^{p\alpha^2 t/2}. \tag{4.2.3}$$

The following lemma will be proved by a similar argument.

Lemma 4.2.4 *Assume that there exists C and $\nu > 0$ such that*

$$\forall t > 0, \quad \|H_t\|_{2\to\infty} \leq (C/t)^{\nu/4}.$$

Then

$$\|H_t^{\alpha, \phi}\|_{p\to q} \leq (C/t)^{\nu(1/p-1/q)/2} e^{q\alpha^2 t/2}$$

for all $q \geq p \geq 2$, $t > 0$, $\alpha \in \mathbb{R}$ and $\phi \in \mathcal{C}_0^\infty(M)$ with $\|\nabla\phi\|_\infty \leq 1$.

Again, consider $H_t^{\alpha z, \phi}$ where α, ϕ, t are fixed, α is real and z is complex with $\Re(z) \in [0,1]$. As above, we have for any $p \in [2, \infty]$,

$$\|H_t^{\alpha(1+bi), \phi}\|_{p\to p} \leq e^{p\alpha^2 t/2}$$

and, by Lemma 4.2.3,

$$\|H_t^{\alpha bi,\phi}\|_{p\to\infty} \le (C/t)^{\nu/2p}.$$

Thus,

$$\|H_t^{\alpha\theta,\phi}\|_{p\to q} \le (C/t)^{\nu(1-\theta)/2p} e^{p\alpha^2\theta t/2}$$

with $1/q = (1-\theta)/\infty + \theta/p$. That is,

$$\|H_t^{\alpha(p/q),\phi}\|_{p\to q} \le (C/t)^{\nu(1/p-1/q)/2} e^{\alpha^2 p^2 t/2q}.$$

Changing α to $p\alpha/q$ yields the desired inequality.

Lemma 4.2.5 *Assume that* $(H_t)_{t>0}$ *satisfies*

$$\forall\, t > 0, \quad \|H_t\|_{2\to\infty} \le (C/t)^{\nu/4}$$

for some $C, \nu > 0$. *For any* $\delta > 0$ *there exists a finite constant* $C(\delta)$ *such that for all* $\alpha \in \mathbb{R}$ *and* $\phi \in \mathcal{C}_0^\infty(M)$ *with* $\|\nabla\phi\|_\infty \le 1$, *we have*

$$\forall\, t > 0, \quad \|H_t^{\alpha,\phi}\|_{2\to\infty} \le (C(\delta)C/t)^{\nu/4}\exp(t\alpha^2(1+\delta)).$$

Let us fix a small $\delta > 0$. Consider two sequences of positive numbers $(s_i)_1^\infty$, $(p_i)_1^\infty$ such that $1 = \sum_1^\infty s_i$, $p_1 = 2$, $p_i \nearrow \infty$ and

$$\sum_1^\infty s_i p_{i+1} \le 2(1+\delta)$$

$$\sum_1^\infty p_i^{-1}\log(1/s_i) = \log[2C(\delta)]$$

with $C(\delta) < \infty$. For instance, take $s_1 = 1 - \epsilon$, $s_i = \epsilon c i^{-5}$ for $i \ge 2$ where $1/c = \sum_2^\infty i^{-5}$, $p_1 = p_2 = 2$, $p_i = 2(i-1)^2$ for $i \ge 2$. Then

$$\sum_1^\infty s_i p_{i+1} = 2(1-\epsilon) + 2\epsilon c \sum_2^\infty i^{-3} \le 2(1+\delta)$$

if $\epsilon > 0$ is chosen so small that $\epsilon c \sum_2^\infty i^{-3} \le \delta$.

Now, using the semigroup property with $t = \sum_1^\infty t_i$, $t_i = ts_i$, write

$$\|H_t^{\alpha,\phi}\|_{2\to\infty} \le \prod_{i=1}^\infty \|H_{t_i}^{\alpha,\phi}\|_{p_i\to p_{i+1}}$$

$$\le \prod_{i=1}^\infty (C/t_i)^{\nu(1/p_i-1/p_{i+1})/2} e^{\alpha^2 p_{i+1} t_i/2}$$

$$\le (C/t)^{\nu/4}\exp\left(t\alpha^2(\sum_1^\infty p_{i+1}s_i/2) + \frac{\nu}{2}(\sum_1^\infty p_i^{-1}\log(1/s_i))\right)$$

$$\le (C(\delta)C/t)^{\nu/4}\exp(t\alpha^2(1+\delta)).$$

This proves Lemma 4.2.5.

4.2.3   Pointwise Gaussian upper bounds

We can now prove the following theorem.

Theorem 4.2.6 *Assume that the manifold M satisfies the Nash inequality*

$$\forall f \in \mathcal{C}_0^\infty(M), \quad \|f\|_2^{2(1+2/\nu)} \le C\|\nabla f\|_2^2\|f\|_1^{4/\nu}.$$

Then, for any $\delta > 0$ there exists a finite constant $C(\delta)$ such that the kernel $h(t,x,y)$ of the heat diffusion semigroup $H_t = e^{-t\Delta}$, $t > 0$, satisfies

$$h(t,x,y) \le (C(\delta)/t)^{\nu/2} \exp\left(-\frac{d(x,y)^2}{4(1+\delta)t}\right).$$

By Theorem 4.1.1, the hypothesis implies that there exists a constant C' such that

$$\forall t > 0, \quad \|H_t\|_{2\to\infty} \le (C/t)^{\nu/4}.$$

Thus, by Lemma 4.2.5, for any $\phi \in \mathcal{C}_0^\infty(M)$ with $\|\nabla\phi\|_\infty \le 1$ and any real α, we have

$$\|H_t^{\alpha,\phi}\|_{2\to\infty} \le (C(\delta)C'/t)^{\nu/4} \exp(t\alpha^2(1+\delta)).$$

As the adjoint of $H_t^{\alpha,\phi}$ is $H_t^{-\alpha,\phi}$ we also have, by duality,

$$\|H_t^{\alpha,\phi}\|_{1\to2} \le (C(\delta)C'/t)^{\nu/4} \exp(t\alpha^2(1+\delta)).$$

Thus

$$\|H_t^{\alpha,\phi}\|_{1\to\infty} \le (2C(\delta)C'/t)^{\nu/2} \exp(t\alpha^2(1+\delta)).$$

That is,

$$h(t,x,y) \le (2C(\delta)C'/t)^{\nu/2} \exp\left(t\alpha^2(1+\delta) + \alpha[\phi(x)-\phi(y)]\right).$$

For fixed x, y, ϕ, take $\alpha = -(\phi(x)-\phi(y))/2(1+\delta)t$. The bound becomes

$$h(t,x,y) \le (2C(\delta)C'/t)^{\nu/2} \exp\left(-\frac{[\phi(x)-\phi(y)]^2}{4(1+\delta)t}\right).$$

We can now optimize over all allowed ϕ's for each fixed $x,y \in M$. The condition $\|\nabla\phi\|_\infty \le 1$ shows that $\phi(x) - \phi(y)$ is at most $d(x,y)$. Moreover, one can find $\phi \in \mathcal{C}_0^\infty(M)$ with $|\nabla\phi| \le 1$ and such that $\phi(x) - \phi(y)$ is as close as we wish to $d(x,y)$. Thus,

$$h(t,x,y) \le (2C(\delta)C'/t)^{\nu/2} \exp\left(-\frac{d(x,y)^2}{4(1+\delta)t}\right)$$

as desired.

One can refine this result as follows. In the proof of Lemma 4.2.5, take $s_1 = 1 - \epsilon$, $s_i = \epsilon c i^{-5}$ for $i \geq 2$ where $1/c = \sum_2^\infty i^{-5}$, $p_1 = p_2 = 2$, $p_i = 2(i-1)^2$ for $i \geq 2$, and take $\epsilon = c_0 \delta$ with

$$c_0 = (c'/c) = \frac{\sum_2^\infty i^{-5}}{\sum_2^\infty i^{-3}}, \quad c' = \frac{1}{\sum_2^\infty i^3}$$

as suggested above. Then $\sum_1^\infty s_i p_{i+1} \leq 2(1+\delta)$. Now, repeating the same argument more carefully, we get

$$\|H_t^{\alpha,\phi}\|_{2\to\infty}$$
$$\leq \prod_{i=1}^\infty \|H_{t_i}^{\alpha,\phi}\|_{p_i \to p_{i+1}}$$
$$\leq \prod_{i=1}^\infty (C'/t_i)^{\nu(1/p_i - 1/p_{i+1})/2} e^{\alpha^2 p_{i+1} t_i / 2}$$
$$\leq (C'/\delta t)^{\nu/4} \exp\left(\frac{t\alpha^2}{2}\left(\sum_1^\infty p_{i+1} s_i\right) + \frac{\nu}{2}\left(\sum_1^\infty p_i^{-1} \log(i^5/c')\right)\right)$$
$$\leq (C''/\delta t)^{\nu/4} \exp(t\alpha^2(1+\delta))$$

where C'' does not depend on δ. This leads to the improved Gaussian upper bound

$$h(t,x,y) \leq (2C''/\delta t)^{\nu/2} \exp\left(-\frac{d(x,y)^2}{4(1+\delta)t}\right).$$

As

$$\frac{d(x,y)^2}{4(1+\delta)t} = \frac{d(x,y)^2}{4t} - \frac{\delta d(x,y)^2}{4(1+\delta)t}$$

picking $\delta = (1 + d(x,y)^2/t)^{-1}$ yields the bound

$$h(t,x,y) \leq 2\left(\frac{2C''}{t}\right)^{\nu/2}\left(1 + \frac{d(x,y)^2}{t}\right)^{\nu/2} \exp\left(-\frac{d(x,y)^2}{4t}\right).$$

This argument also gives Gaussian upper bounds for the time derivatives of $h(t,x,y)$. Namely, using (4.1.3) and Lemma 4.2.2, we obtain that under the assumption of Theorem 4.2.6,

$$|\partial_t^k h(t,x,y)| \leq C_k t^{-\nu/2-k}\left(1 + \frac{d(x,y)^2}{t}\right)^{\nu/2+k} \exp\left(-\frac{d(x,y)^2}{4t}\right). \quad (4.2.4)$$

4.2.4 On-diagonal lower bounds

Here we want to discuss briefly an application of the Gaussian upper bounds derived above. Let M be a Riemannian manifold of dimension n satisfying the Nash inequality

$$\forall f \in \mathcal{C}_0^\infty(M), \quad \|f\|_2^{2(1+2/\nu)} \leq C\|\nabla f\|_2^2 \|f\|_1^{4/\nu}$$

for some $\nu > 0$. This implies the volume lower bound

$$\forall t > 0, \quad V(t) \geq ct^{\nu}$$

which in turn implies $\nu \geq n$. If all we have is this information, the Gaussian upper bound of Theorem 4.2.6 is not extremely useful. In particular, if the volume growth is actually faster than t^{ν} (e.g., t^a with $a > \nu$, $t > 1$), this bound is hard to use. This can be seen from the following elementary lemma.

Lemma 4.2.7 *Assume that M has volume growth $V(x,t)$ with*

$$\forall x \in M, \ \forall t \geq 1, \ c_0 \leq t^{-a} V(x,t) \leq C_0.$$

Then, for all $x, y \in M$, $t > 0$, $R \geq 1$,

$$ct^{a/2} \exp(-CR^2/t) \leq \int_{d(x,y) \geq R} \exp(-d(x,y)^2/t) dy \leq Ct^{a/2} \exp(-cR^2/t).$$

Write

$$I(t,R) = \int_{d(x,y) \geq R} e^{-d(x,y)^2/t} dy = \sum_0^{\infty} \int_{R2^k \leq d(x,y) \leq R2^{k+1}} e^{-d(x,y)^2/t} dy.$$

Thus

$$\begin{aligned} I(t,R) &\leq C_0 R^a \sum_0^{\infty} 2^{(k+1)a} e^{-R^2 2^{2k}/t} \\[2mm] &\leq C_0 R^a \left(\sum_{k: R^2 2^{2k} \leq t} 2^{(k+1)a} e^{-R^2 2^{2k}/t} \right. \\[2mm] &\qquad \left. + \sum_{k: R^2 2^{2k} > t} 2^{(k+1)a} e^{-R^2 2^{2k}/t} \right). \end{aligned}$$

Let $k(t)$ be the smallest integer k such that $R^2 2^{2k} > t$ (such an integer always exists). Then $R^2 2^{2k} \leq t/4$ for $k < k(t)$. On the one hand,

$$\sum_{k < k(t)} 2^{(k+1)a} e^{-R^2 2^{2k}/t} \leq C_a 2^{ak(t)}$$

and

$$C_0 R^a \sum_{k < k(t)} 2^{(k+1)a} e^{-R^2 2^{2k}/t} \leq C_0 C_a R^a 2^{ak(t)} \leq C_a' t^{a/2}.$$

On the other hand,

$$\sum_{k \geq k(t)} 2^{(k+1)a} e^{-R^2 2^{2k}/t} \leq e^{-1} 2^a e^{-R^2/t} 2^{ak(t)} \sum_{k \geq k(t)} 2^{a(k-k(t))} e^{-2^{2(k-k(t))}}$$

from which it follows that

$$C_0 R^a \sum_{k \geq k(t)} 2^{(k+1)a} e^{-R^2 2^{2k}/t} \leq C_a'' t^{a/2} e^{-R^2/2t}.$$

This proves the upper bound. For the lower bound, it suffices to restrict the integral to a ball around a point z at distance $2R + \sqrt{t}$ from x and of radius $R + \sqrt{t}$.

Lemma 4.2.7 shows that, on the one hand, the integral of the Gaussian upper bound of Theorem 4.2.6 over the complement of the ball of radius R is not uniformly bounded as $t \to \infty$ when $V(x,t) \simeq t^a$ with $a > \nu$. This should be compared with the fact that $\int_M h(t,x,y)dy = 1$! On the other hand, if $V(x,t) \simeq t^\nu$, then we have

$$\int_{d(x,y) \geq R} h(t,x,y)dy \leq C e^{-cR^2/t} \qquad (4.2.5)$$

for all $t > 0$, $R \geq 1$, which is an informative result.

Our aim in this section is to prove the following theorem.

Theorem 4.2.8 *Fix $\nu > 0$. Assume that M satisfies the Nash inequality*

$$\forall f \in C_0^\infty(M), \quad \|f\|_2^{2(1+2/\nu)} \leq C\|\nabla f\|_2^2 \|f\|_1^{4/\nu}$$

and the volume growth condition

$$\forall x \in M, \ \forall r > 0, \quad c_0 \leq r^{-\nu} V(x,r) \leq C_0.$$

Then the heat kernel $h(t,x,y)$ is bounded above and below on the diagonal by

$$ct^{-\nu/2} \leq h(t,x,x) \leq Ct^{-\nu/2}.$$

The proof uses the fact that, under the above hypotheses, $\int_M h(t,x,y)dy = 1$. This is not obvious and requires a proof, which can be found in Section 5.5.2. Assuming that indeed $\int_M h(t,x,y)dy = 1$, (4.2.5) implies

$$\int_{B(x,A\sqrt{t})} h(t,x,y)dy \geq 1/2$$

for all $t > 1$ and A large enough. By Jensen's inequality, this yields

$$
\begin{aligned}
h(2t,x,x) &= \int h(t,x,y)^2 dy \\
&\geq \int_{B(x,A\sqrt{t})} h(t,x,y)^2 dy \\
&\geq V(x,A\sqrt{t})^{-1} \int_{B(x,A\sqrt{t})} h(t,x,y)dy \\
&\geq \frac{1}{2V(x,A\sqrt{t})}.
\end{aligned}
$$

The theorem follows.

We now give a quite different proof of a weaker lower bound, namely

$$\sup_{x \in M} h(t, x, x) \geq ct^{-\nu/2}.$$

This lower bound is taken from [18]. It is weaker because only the supremum of $h(t, x, x)$ is bounded from below. On the other hand it requires no assumption except the volume estimate

$$\forall x \in M, \ \forall r > 0, \quad c_0 \leq r^{-\nu} V(x, r) \leq C_0.$$

No Nash inequality is required and the proof does not assume that

$$\int_M h(t, x, y) dy = 1$$

although this is true under the above volume hypothesis.

We start with the following computation. Fix $f \in L^2(M)$ with $\|f\|_2 = 1$. Then one can check that

$$\frac{\partial_t \|H_t f\|_2^2}{\|H_t f\|_2^2} \geq -2\|\nabla f\|_2^2. \tag{4.2.6}$$

Indeed, the right-hand side is the value of the left-hand side at $t = 0$. Computing the derivative of the left-hand side, one gets

$$\frac{(\partial_t^2 \|H_t f\|_2^2) \|H_t f\|_2^2 - (\partial_t \|H_t f\|_2^2)^2}{\|H_t f\|_2^4}.$$

The numerator is equal to

$$4 \left(\langle \Delta^2 H_t f, H_t f \rangle \langle H_t f, H_t f \rangle - \langle \Delta H_t f, H_t f \rangle^2 \right)$$

which is positive by the Cauchy–Schwarz inequality. This proves (4.2.6).

Now, (4.2.6) implies

$$\exp \left(-2\|\nabla f\|_2^2 \, t \right) \leq \|H_t f\|_2^2.$$

As

$$\sup_{x \in M} h(t, x, x) = \sup_{x \in M} \|h(t/2, x, \cdot)\|_2^2 = \|H_t\|_{2 \to \infty}^2 = \|H_{t/2}\|_{1 \to 2}^2,$$

we get, for any function $f \in C_0^\infty(M)$,

$$\sup_{x \in M} h(t, x, x) \geq \frac{\|f\|_2^2}{\|f\|_1^2} \exp \left(-2 \frac{\|\nabla f\|_2^2}{\|f\|_2^2} \, t \right). \tag{4.2.7}$$

For any fixed t, consider a function f supported in a ball $B = B(o, 2\sqrt{t})$ for some fixed $o \in M$. Then, by Jensen's inequality $\|f\|_2^2 / \|f\|_1^2 \geq V(o, 2\sqrt{t})^{-1}$. Specialize f to be equal to $f_t(y) = (d(o, y) - \sqrt{t})_+$. Then

$$\|\nabla f_t\|_2^2 \leq V(o, 2\sqrt{t}), \quad \|f_t\|_2^2 \geq tV(o, \sqrt{t}).$$

Hence,

$$\sup_{x \in M}\{h(t,x,x)\} \geq \frac{1}{V(o,2\sqrt{t})} \exp\left(-\frac{V(o,2\sqrt{t})}{V(o,\sqrt{t})}\right).$$

Now, the desired inequality follows from the fact that $V(o,r) \simeq r^\nu$. In fact, this proves the bound

$$\sup_{x} h(t,x,x) \geq c\frac{1}{V(o,\sqrt{t})}$$

under the sole hypothesis that

$$\frac{V(o,2\sqrt{t})}{V(o,\sqrt{t})}$$

is bounded above by a constant independent of $t > 0$.

4.3 The Rozenblum–Lieb–Cwikel inequality

4.3.1 The Schrödinger operator $\Delta - V$

Let M be a complete non-compact Riemannian manifold with Laplace–Beltrami operator

$$\Delta f = -\mathrm{div}(\nabla f).$$

Recall that our convention is that Δ is a positive operator in the L^2 sense, that is, $\langle \Delta f, f \rangle \geq 0$ for all $f \in C_0^\infty(M)$. In other words, the spectrum of Δ is contained in the non-negative semi-axis $[0,\infty)$. We will denote by dx the Riemannian measure on M and often write

$$\int_M f(x)dx = \int_M f.$$

Consider now the Schrödinger operator

$$L = \Delta - V$$

where V is a non-negative function. Then L may well have some negative spectrum. However, if V is a nice bounded function vanishing at infinity, one may hope that the essential spectra of L and Δ coincide. In this case, the negative spectrum of L is a discrete set with, possibly, an accumulation point at 0 (if 0 is indeed the bottom of the spectrum of Δ). A natural question is: what condition on V will imply a bound on the number of negative eigenvalues of L?

The following result (for $M = \mathbb{R}^n$, $n \geq 3$) was first proved by Rozenblum and is known as the Rozenblum–Lieb–Cwikel inequality.

Theorem 4.3.1 *Let M be a Riemannian manifold satisfying an (L^2, ν)-Sobolev inequality for some $\nu > 2$. Let $L = \Delta - V$ with $V \in L^1_{\text{loc}}(M)$ and $V_+ \in L^{\nu/2}$. Let $\mathcal{N}_V(\lambda)$ be the number of eigenvalues of L less than λ, counting multiplicity. Then*

$$\mathcal{N}_V(0) \leq C(\nu) \int_M V_+^{\nu/2}. \tag{4.3.1}$$

This bound does not hold, in general, for the number of eigenvalues less than or equal to 0. We will follow a proof due to Li and Yau [55] which is also presented in a more general setting in [52]. See also [56] and the references given in [52].

Before embarking on the proof, let us observe that, if we assume that inequality (4.3.1) holds for all potentials V in $\mathcal{C}_0^\infty(M)$, then L is a nonnegative operator for all $V \in \mathcal{C}_0^\infty(M)$ with

$$\|V\|_{\nu/2} \leq C(\nu)^{-1}.$$

That is,

$$0 \leq \int |\nabla f|^2 - \int V|f|^2$$

for all such V and all $f \in \mathcal{C}_0^\infty(M)$, i.e.,

$$\sup_{\substack{V \in \mathcal{C}_0^\infty(M) \\ \|V\|_{\nu/2} \leq 1/C(\nu)}} \left\{ \int V|f|^2 \right\} \leq \int |\nabla f|^2.$$

As the dual of $L^{\nu/2}$ is $L^{\nu/(\nu-2)}$, this yields

$$\forall f \in \mathcal{C}_0^\infty(M), \quad \left(\int |f|^{2\nu/(\nu-2)} \right)^{(\nu-2)/\nu} \leq C(\nu) \int |\nabla f|^2.$$

In words, the Rozenblum–Lieb–Cwikel inequality (4.3.1) with parameter $\nu > 2$ easily implies that M satisfies an (L^2, ν)-Sobolev inequality. According to Theorem 4.3.1, the Rozenblum–Lieb–Cwikel inequality (4.3.1) can thus be seen as yet another form —the strongest form, in some sense— of the Sobolev inequality

$$\|f\|_{2\nu/(\nu-2)}^2 \leq C \|\nabla f\|_2^2.$$

Let us note here that the problem of computing the best constant $C(\nu)$ for the Rozenblum–Lieb–Cwikel inequality (4.3.1) in $\mathbb{R}^n$, $n = \nu$, is an important open problem (e.g., $n = 3$). It is known that this best constant is strictly larger than the best constant in the corresponding Sobolev inequality. The smallest known constant $C(\nu)$ in (4.3.1) was obtained by Lieb [56]. See also [73].

4.3.2   The operator $T_V = \Delta^{-1}V$

Assume that M satisfies an (L^2, ν)-Sobolev inequality with $\nu > 2$ and set $q = 2\nu/(\nu - 2)$. Thus

$$\forall f \in \mathcal{C}_0^\infty(M), \quad \left(\int_M |f|^q\right)^{2/q} \leq C \int_M |\nabla f|^2. \qquad (4.3.2)$$

Then $\|\nabla f\|_2$ is a norm on $\mathcal{C}_0^\infty(M)$, and we can take the completion of $\mathcal{C}_0^\infty(M)$ for this norm. We obtain a Hilbert space H. According to (4.3.2), the map $J : \mathcal{C}_0^\infty(M) \to L^q$ extends as a continuous map from H to L^q. We want to show that J is one to one, that is, that H can be viewed as a closed subspace of L^q with norm $\|\nabla f\|_2$. That this is the case does not immediately follow from (4.3.2). Let $F = (f_n)$ be a Cauchy sequence (i.e., an element of H) for $g \to \|\nabla g\|_2$ on $\mathcal{C}_0^\infty(M)$. Note that (∇f_n) converges in L^2 to a certain vector field X. By (4.3.2), there exists $f \in L^q$ such that $f_n \to f$ in L^q. Let Ω be any bounded domain. As $L^q(\Omega) \subset L^2(\Omega)$ for $q > 2$ and since the restriction of $g \to \|\nabla g\|_2^2$ to any bounded domain defines a *closed* form, the restrictions of the f_n's to Ω converge to the restriction of f to Ω *in the norm* $g \to \|\nabla g\|_{2,\Omega} + \|g\|_{2,\Omega}$. If $f = 0$, we must have $X = \nabla f = 0$ in any bounded domain Ω. Thus $\lim \|\nabla f_n\|_2 = 0$, that is $F = (f_n) = 0$ in H.

Fix $V \in L^{\nu/2}$. Then, by Hölder's inequality with conjugate exponents

$$\frac{\nu}{\nu - 2}, \quad \frac{\nu}{2}$$

and the (L^2, ν)-Sobolev inequality,

$$\begin{aligned}
\left|\int |f|^2 V\right| &\leq \|f^2\|_{\nu/(\nu-2)}\|V\|_{\nu/2} \\
&= \|f\|_{2\nu/(\nu-2)}^2\|V\|_{\nu/2} \\
&\leq C\|V\|_{\nu/2}\|\nabla f\|_2^2.
\end{aligned} \qquad (4.3.3)$$

It follows that the self-adjoint operator T_V associated to the quadratic form $\int_M V|f|^2$ on H is a bounded operator on H. The action of T_V on $\mathcal{C}_0^\infty(M) \subset H$ can easily be identified if we assume that V is smooth. Indeed, we then have

$$\begin{aligned}
\int (Vf)g &= \int (\Delta\Delta^{-1}Vf)g \\
&= \int \langle \nabla(\Delta^{-1}Vf), \nabla g \rangle
\end{aligned}$$

for all $f, g \in \mathcal{C}_0^\infty(M)$. That is,

$$T_V = \Delta^{-1}V. \qquad (4.3.4)$$

Theorem 4.3.2 *Assume that M satisfies the Sobolev inequality* (4.3.2) *with $\nu > 2$. Fix $V \in L^{\nu/2}$. Then $T_V : H \rightarrow H$ is a compact operator and $\mathbf{N}_{T_V}(\lambda) = \#\{k : \lambda_k > \lambda\}$, the number of eigenvalues of T_V larger than λ (counting multiplicity), is bounded by*

$$\mathbf{N}_{T_V}(\lambda) \leq (eC/\lambda)^{\nu/2} \int_M V_+^{\nu/2}.$$

The first step in the proof of this result is to reduce it to the case where $V \in L^1 \cap L^\infty$, V is smooth and $V > 0$. We will not do this in detail. However, this reduction is quite easy once one interprets the desired inequality as a boundedness inequality for the linear map $V \mapsto T_V$ between $L^{\nu/2}$ and the appropriate Banach space of compact operators. Thus, from now on, we assume that $V \in L^1 \cap L^\infty$, V is smooth and $V > 0$.

Let μ denote the measure

$$d\mu(x) = V(x)dx$$

on M. Consider the space $L^2(M, \mu)$. By (4.3.3), this space contains H. Consider the quadratic form $\|\nabla f\|_2^2$ on $L^2(M, d\mu)$, with domain H. By (4.3.3) and the definition of H, this form is positive definite and closed. Thus, it is a Dirichlet form on $L^2(M, d\mu)$. Actually, it is easy to compute the associated infinitesimal generator $-K$. Indeed,

$$\int \langle \nabla f, \nabla g \rangle = \int (\Delta f)g = \int (V^{-1}\Delta f)g d\mu.$$

Hence $K = V^{-1}\Delta f$. By (4.3.4), this means that

$$K^{-1}f = T_V f \tag{4.3.5}$$

on $\mathcal{C}_0^\infty(M)$ (which is dense in the domain H of T_V!).

As V is bounded and M satisfies (4.3.2), we have

$$\left(\int |f|^q d\mu \right)^{2/q} \leq C \|V\|_\infty^{2/q} \int |\nabla f|^2$$

where $q = 2\nu/(\nu - 2)$. That is, the Dirichlet form $Q(f, f) = \int |\nabla f|^2$ on $L^2(M, d\mu)$ satisfies a Sobolev inequality. As Sobolev inequality implies Nash inequality, we can apply Theorem 4.1.1 to Q. This shows that the semigroup $K_t = e^{-tK}$ is ultracontractive. In particular, for each $t > 0$, K_t has a bounded kernel $k(t, x, y)$ w.r.t. $d\mu$. As $\int_M d\mu$ is finite, it follows that the function k defined by

$$k(t) = \int k(t, x, y)^2 d\mu(x)d\mu(y)$$

is finite. That is, K_t is a Hilbert–Schmidt operator. Therefore, the spectrum of the self-adjoint operator K is discrete. Let λ_i, $i = 0, 1, 2, \ldots$ be the eigenvalues of K repeated according to multiplicity and in non-decreasing order. Let u_i be the corresponding real eigenfunctions normalized in $L^2(M, d\mu)$. The kernel $k(t, x, y)$ can be expressed in these terms as

$$k(t, x, y) = \sum_i e^{-t\lambda_i} u_i(x) u_i(y).$$

Hence

$$k(t) = \sum_i e^{-2\lambda_i t}.$$

By (4.3.5),

$$\mathbf{N}_{T_V}(\lambda) = \#\{i : \lambda_i < \lambda^{-1}\}$$

and thus

$$\mathbf{N}_{T_V}(\lambda) \leq \sum_{i : \lambda_i \lambda < 1} e^{2(1 - \lambda_i \lambda)t}$$

$$\leq e^{2t} k(\lambda t) \qquad (4.3.6)$$

for all $\lambda, t > 0$. This reduces Theorem 4.3.2 to a suitable bound on $k(t)$ as function of t, which is given by the following lemma.

Lemma 4.3.3 *Assume that $\|V\|_{\nu/2} = 1$. Then the function $k(t)$ satisfies*

$$k(t) \leq (-Ck'(t)/2)^{\nu/(\nu+2)}.$$

In particular

$$k(t) \leq (C\nu/4t)^{\nu/2}.$$

Here C is the constant appearing in (4.3.2).

Using this in (4.3.6) with $t = \nu/4$ yields

Corollary 4.3.4 *Assume that $\|V\|_{\nu/2} = 1$. Then*

$$\mathbf{N}_{T_V}(\lambda) \leq (eC/\lambda)^{\nu/2}.$$

Let us prove Lemma 4.3.3. Observe that

$$2 = \frac{2\nu}{\nu + 2} + \frac{4}{\nu + 2}$$

and that the Hölder conjugate of $4/(\nu + 2)$ is $(\nu - 2)/(\nu + 2)$. Then write

$$\int k(t, x, y)^2 V(y) \, dy$$

$$= \int k(t, x, y)^{2\nu/(\nu+2)} [k(t, x, y) V(y)^{(\nu+2)/4}]^{4/(\nu+2)} \, dy$$

$$\leq \left(\int k(t, x, y)^{2\nu/(\nu-2)} \, dy \right)^{(\nu-2)/(\nu+2)} \left(\int k(t, x, y) V(y)^{(\nu+2)/4} \, dy \right)^{4/(\nu+2)}$$

$$= \left(\int k(t, x, y)^q \, dy \right)^{2\nu/q(\nu+2)} \left(K_t[V^{(\nu-2)/4}](x) \right)^{4/(\nu+2)}.$$

Note that the last factor is indeed

$$K_t[V^{(\nu-2)/4}](x) \ = \ \int k(t,x,y)V(y)^{(\nu-2)/4}V(y)dy$$

$$= \ \int k(t,x,y)V(y)^{(\nu+2)/4}dy.$$

Let us consider the factor $(\int k(t,x,y)^q dy)^{2/q}$. We want to estimate this factor by using the Sobolev inequality (4.3.2), that is, we want to write

$$\left(\int k(t,x,y)^q dy\right)^{1/q} \leq C \int |\nabla_y k_t(x,y)|^2 dy = CQ(k_t^x, k_t^x)$$

where $k_t^x(y) = k(t,x,y)$. This is legitimate if we can show that $k_t^x \in H$. Fortunately, this easily follows from the facts that $K_t f \in H$ for all $t > 0$ and all $f \in L^2(M,d\mu)$ and that $k_t^x \in L^2(M,d\mu)$, $t > 0$. The latter is a consequence of the ultracontractivity of K_t since $k(2t,x,x) = \|k_t^x\|_{L^2(M,d\mu)}^2$. The former is a general fact about analytic semigroups of operators (any self-adjoint (C_0) semigroup of contractions on $L^2(M,d\mu)$ is analytic). See Section 4.1.1 and (4.1.2), (4.1.4). Thus

$$\int k(t,x,y)^2 V(y)dy \leq [CQ(k_t^x,k_t^x)]^{\nu/(\nu+2)} \left(K_t[V^{(\nu-2)/4}](x)\right)^{4/(\nu+2)}. \quad (4.3.7)$$

Observe also that

$$Q(k_t^x, k_t^x) = -\int (\partial_t k_t^x(y))k_t^x(y)V(y)dy$$

so that

$$\int Q(k_t^x, k_t^x)V(x)dx \ = \ -\int (\partial_t k(t,x,y))k(t,x,y)V(x)V(y)dxdy$$

$$= \ -\frac{1}{2}\partial_t \int k(t,x,y)^2 V(x)V(y)dxdy$$

$$= \ -\frac{1}{2}k'(t).$$

The change in the order of derivative and integral is easily justified by the ultracontractivity of K_t and (4.1.4) (i.e., the analyticity of K_t on $L^2(M,d\mu)$).

We now integrate (4.3.7) against $d\mu(x) = V(x)dx$ and use that $(\nu+2)/\nu$ and $(\nu+2)/2$ are Hölder conjugate exponents to get

$$k(t) \ = \ \int k(t,x,y)^2 V(y)V(x)dydx$$

$$\leq \ \int [CQ(k_t^x,k_t^x)V(x)dx]^{\nu/(\nu+2)} \left(K_t[V^{(\nu-2)/4}](x)\right)^{4/(\nu+2)} V(x)dx$$

$$\leq \left(C \int Q(k_t^x, k_t^x) V(x) dx V(y) dy\right)^{\nu/(\nu+2)}$$

$$\times \left(\int \left(K_t[V^{(\nu-2)/4}](x)\right)^2 V(x) dx\right)^{4/(\nu+2)}$$

$$\leq [-(C/2)k'(t)]^{\nu/(\nu+2)} \left(\int (K_t[V^{(\nu-2)/4}](x))^2 V(x) dx\right)^{2/(\nu+2)}$$

$$\leq [-(C/2)k'(t)]^{\nu/(\nu+2)} \left(\int V(x)^{(\nu-1)/2} V(x) dx\right)^{2/(\nu+2)}$$

$$\leq [-(C/2)k'(t)]^{\nu/(\nu+2)} \left(\int V(x)^{\nu/2} dx\right)^{2/(\nu+2)}$$

$$\leq [-(C/2)k'(t)]^{\nu/(\nu+2)}.$$

Here, we have used that K_t is a contraction on $L^p(M, d\mu)$, for all $1 \leq p \leq \infty$, in particular for $p = 2$, and also

$$\int V(x)^{(\nu-1)/2} V(x) dx = \|V\|_{\nu/2}^{\nu/2} = 1.$$

This proves Lemma 4.3.3.

4.3.3 The Birman–Schwinger principle

That the estimate of Theorem 4.3.2 is equivalent to the estimate of Theorem 4.3.1 is known as the Birman–Schwinger principle. More precisely, for $V \geq 0$, we have

$$\mathbf{N}_{T_V}(1) = \mathcal{N}_V(0) \tag{4.3.8}$$

where $\mathcal{N}_V(0)$ is the number of negative eigenvalues of $\Delta - V$. To see this observe that the Rayleigh ratio for $\Delta - V$ on $L^2(M, dx)$ is

$$R_L(f) = \frac{\int (|\nabla f(x)|^2 - V(x)|f(x)|^2) dx}{\int |f(x)|^2 dx}$$

whereas the Rayleigh ratio for K on $L^2(M, d\mu)$, $d\mu(x) = V(x) dx$, is

$$R_K(f) = \frac{\int |\nabla f(x)|^2 dx}{\int |f(x)|^2 V(x) dx}$$

and

$$\frac{\int (|\nabla f(x)|^2 - V(x)|f(x)|^2) dx}{\int |f(x)|^2 dx} = \frac{\int |f(x)|^2 V(x) dx}{\int |f(x)|^2 dx} \left(\frac{\int |\nabla f(x)|^2 dx}{\int |f(x)|^2 V(x) dx} - 1\right).$$

It follows that the number of eigenvalues of K less than 1, that is $\mathbf{N}_{T_V}(1)$, is equal to the number of negative eigenvalues of $\Delta - V$, that is $\mathcal{N}_V(0)$. Indeed, a classical argument shows that

$$\mathbf{N}_{T_V}(\lambda) = \sup\{ \dim(F) : F \subset \mathcal{C}_0^\infty(M), R_K(f) > \lambda, 0 \neq f \in \mathcal{C}_0^\infty(M)\}$$

and

$$\mathcal{N}_V(\lambda) = \sup\{\ \dim(F) : F \subset \mathcal{C}_0^\infty(M), R_L(f) < \lambda, 0 \neq f \in \mathcal{C}_0^\infty(M)\}.$$

Note that, because we are working with functions in $\mathcal{C}_0^\infty(M)$ on the right-hand side, the above formulas only hold for counting eigenvalues strictly larger (or smaller) than a given value λ.

This proves (4.3.8) and ends the proof of the Rozenblum–Lieb–Cwikel inequality, i.e., Theorem 4.3.1.

Chapter 5

Parabolic Harnack inequalities

The aim of this chapter is to characterize those manifolds that satisfy a scale-invariant parabolic Harnack principle. This is achieved in Section 5.5.

5.1 Scale-invariant Harnack principle

Let us start by considering the elliptic version of the Harnack principle since it is easier to grasp. Given a complete Riemannian manifold M, we say that M satisfies a scale-invariant elliptic Harnack principle if there exists a constant C such that, for any ball B in M and any positive solution u of $\Delta u = 0$ in B, we have

$$\sup_{\frac{1}{2}B}\{u\} \leq C \inf_{\frac{1}{2}B}\{u\}.$$

Here $\frac{1}{2}B$ is the ball with the same center as B and radius half that of B. Note that this inequality is uniform with respect to: (1) the center of B, (2) the radius of B, (3) the harmonic function u.

We now describe the parabolic version of this scale-invariant Harnack principle. For any $x \in M$ and $s \in \mathbb{R}$, $r > 0$, let $Q = Q(x, s, r)$ be the cylinder

$$Q(x, s, r) = (s - r^2, s) \times B(x, r).$$

Let Q_+, Q_- be respectively the upper and lower sub-cylinders

$$
\begin{aligned}
Q_+ &= (s - (1/4)r^2, s) \times B(x, (1/2)r) \\
Q_- &= (s - (3/4)r^2, s - (1/2)r^2) \times B(x, (1/2)r).
\end{aligned}
$$

We say that M satisfies a scale-invariant parabolic Harnack principle if there exists a constant C such that for any $x \in M$, $s \in \mathbb{R}$, $r > 0$ and any positive solution u of $(\partial_t + \Delta)u = 0$ in $Q = Q(x, s, r)$ we have

$$\sup_{Q_-}\{u\} \leq C \inf_{Q_+}\{u\}.$$

Thus, the main difference between the elliptic and parabolic cases is that the cylinders Q_-, Q_+ are disjoint whereas, in the elliptic case, the sup and inf are taken over the same ball $\frac{1}{2}B$. It is easy to see the necessity of the lapse of time separating Q_- from Q_+: consider the fundamental solution

$$u_y(t,x) = \left(\frac{1}{4\pi t}\right)^{n/2} \exp\left(-\frac{|x-y|^2}{4t}\right)$$

of the heat equation on $\mathbb{R}^n$. Letting $\|y\| = \tau$, one easily computes the sup and inf of u over the ball $B = B(0,r)$ with $r < \tau$. One finds that

$$\frac{\sup_B\{u_y(t,\cdot)\}}{\inf_B\{u_y(t,\cdot)\}} = \exp\left(\frac{1}{4t}[(r+\tau)^2 - (r-\tau)^2]\right) = \exp\left(\frac{\tau r}{t}\right).$$

Taking $r = \sqrt{t}$ and $\tau \gg \sqrt{t}$ one sees that this ratio is unbounded as t, τ tend to infinity.

It turns out that this scale-invariant Harnack principle contains a lot of information. To give only one of the striking consequences that will be developed below, the scale-invariant Harnack principle implies that the fundamental solution $h(t,x,y)$ of the heat diffusion equation $(\partial_t + \Delta)u = 0$ satisfies the two-sided Gaussian estimate

$$\frac{c_1}{V(x,\sqrt{t})}e^{-C_1 d(x,y)^2/t} \le h(t,x,y) \le \frac{C_2}{V(x,\sqrt{t})}e^{-c_2 d(x,y)^2/t}.$$

In fact this two-sided bound is equivalent to the parabolic Harnack principle and holds if and only if the following two properties are satisfied:

- there exists a constant D_0 such that the volume growth function $V(x,r)$ has the doubling property

$$\forall\, x \in M, \ \forall\, r > 0, \quad V(x,2r) \le D_0 V(x,r);$$

- there exists a constant P_0 such that the Poincaré inequality

$$\forall\, x \in M, \ \forall\, r > 0, \quad \int_{B(x,r)} |f - f_B|^2 d\mu \le P_0 r^2 \int_{B(x,2r)} |\nabla f|^2 d\mu$$

is satisfied.

That these two properties are sufficient to imply the scale-invariant parabolic Harnack principle was proved independently by A. Grigor'yan [32] and the author [74] by two different methods. The necessity is proved in [74], building on an idea of Kusuoka and Stroock [50]. See the discussion at the beginning of Section 5.5. In order to study parabolic Harnack inequalities we will have to develop a number of techniques which are of interest in their own right.

The parabolic Harnack principle above follows easily from the type of gradient estimates obtained by P. Li and S-T. Yau in [55] under Ricci curvature lower bounds; see also the treatment in [21]. The approach developed below is very different in spirit and can be used to treat cases where no gradient bound can possibly hold, e.g., in the case of equations having non-smooth coefficients or manifolds with non-smooth metrics. Moreover it seems to be a difficult problem to characterize manifolds on which a uniform gradient Harnack estimate holds true whereas we will be able to characterize those manifolds satisfying a parabolic Harnack inequality.

5.2 Local Sobolev inequalities

5.2.1 Local Sobolev inequalities and volume growth

Recall that the validity of a Sobolev inequality such as

$$\forall f \in \mathcal{C}_0^\infty(M), \quad \|f\|_q \leq C\|\nabla f\|_2$$

with $q = 2\nu/(\nu - 2)$ implies the volume lower bound

$$V(x, r) \geq cr^\nu$$

for all $r > 0$. This rules out many interesting manifolds for which one would expect to be able to produce a global analysis of, say, the heat kernel. For instance, the Euclidean cylinders $\mathbb{R}^n \times \mathbb{T}^m$, $n, m < \infty$, cannot carry a Sobolev inequality of the type above, for any ν.

It turns out that there is a simple way to deal with this difficulty. It consists in working with localized Sobolev inequalities.

Fix $1 \leq p < \infty$. Very generally, we say that a Riemannian manifold M satisfies a (family of) localized L^p Sobolev inequality(ies) with constants $C(B)$ and exponent $\nu > p$ if, for any geodesic ball B and all $f \in \mathcal{C}_0^\infty(B)$,

$$\left(\int |f|^q d\mu\right)^{p/q} \leq C(B)\frac{r(B)^p}{\mu(B)^{p/\nu}} \int (|\nabla f|^p + r(B)^{-p}|f|^p)d\mu \qquad (5.2.1)$$

where $q = p\nu/(\nu - p)$, $r(B)$ is the radius of B and $\mu(B)$ is its volume. It is not hard to see that any complete non-compact manifold M satisfies such a family of Sobolev inequalities. The essential information is then concentrated in the behavior of the constant $C(B)$. The precise value of the parameter $\nu > p$ plays only a minor role. In fact, the inequality above can be written

$$\left(\frac{1}{\mu(B)}\int |f|^q d\mu\right)^{p/q} \leq C(B)r(B)^p\left(\frac{1}{\mu(B)}\int (|\nabla f|^p + r(B)^{-p}|f|^p)d\mu\right)$$

which shows that one can always freely increase the value of ν, that is, decrease the value of $q > p$. It will turn out that increasing ν will have little impact on the conclusions that we will draw from these inequalities.

One of the most interesting cases is when M satisfies a family of localized Sobolev inequalities with $\sup_B C(B) = C < \infty$. This condition implies a different kind of control of the volume growth function. Indeed, the argument of Theorem 3.1.5 easily yields the following two results.

Theorem 5.2.1 *Let B be a fixed ball of radius $r(B)$ in M. Assume that the Sobolev inequality*

$$\left(\int |f|^{p\nu/(\nu-p)} d\mu \right)^{(\nu-p)/\nu} \le C(B) \frac{r(B)^p}{\mu(B)^{p/\nu}} \int (|\nabla f|^p + r(B)^{-p}|f|^p) d\mu$$

holds in this ball for some $\nu > p$ and all $f \in \mathcal{C}_0^\infty(B)$. Then there exists a constant C_1 such that

$$\frac{\mu(B)}{\mu(B')} \le C_1 \left(\frac{r(B)}{r(B')} \right)^\nu$$

for all balls $B' \subset B$.

Theorem 5.2.2 *Assume that M satisfies (5.2.1) with $\sup_B C(B) = C < \infty$ and some $\nu > p$. Then there exists a constant C_1 such that*

$$\frac{V(x,T)}{V(x,t)} \le C_1 \left(\frac{T}{t} \right)^\nu$$

for all $x \in M$ and $0 < t < T < \infty$. In particular, M satisfies the doubling condition

$$\forall x \in M, \ \forall t > 0, \quad V(x,2t) \le D_0 V(x,t).$$

The next theorem gives a kind of converse under the additional crucial hypothesis that M satisfies a pseudo-Poincaré inequality.

Theorem 5.2.3 *Assume that M is complete, not compact, and satisfies the pseudo-Poincaré inequality*

$$\forall f \in \mathcal{C}_0^\infty(M), \ \forall s > 0, \quad \|f - f_s\|_p \le C_0 \, s \, \|\nabla f\|_p$$

where $f_s(x) = V(x,s)^{-1} \int_{B(x,s)} f d\mu$. Assume also that M satisfies the doubling volume growth condition

$$\forall x \in M, \ \forall t > 0, \quad V(x,2t) \le D_0 V(x,t).$$

Then there exists a real $\nu > p$ such that (5.2.1) holds true with $\sup_B C(B) = C < \infty$, that is, for any ball B,

$$\forall f \in \mathcal{C}_0^\infty(B), \quad \left(\int |f|^{p\nu/(\nu-p)} d\mu \right)^{(\nu-p)/\nu} \le C \frac{r(B)^p}{\mu(B)^{p/\nu}} \int |\nabla f|^p d\mu.$$

The proof starts with the following two easy lemmas.

Lemma 5.2.4 *If M satisfies the doubling volume growth condition*

$$\forall\, x \in M, \ \ \forall\, t > 0, \ \ V(x, 2t) \leq D_0 V(x, t)$$

then

$$\frac{V(x, T)}{V(y, t)} \leq D_0^2 \left(\frac{T}{t}\right)^{\nu_0}$$

with $\nu_0 = \log_2(D_0)$ for all $0 < t < T < \infty$ and all $x \in M$, $y \in B(x, T)$.

Let k be such that $2^k < T/t \leq 2^{k+1}$. As $B(x, T) \subset B(y, 2T) \subset B(y, 2^{k+2}t)$ and thus $V(x, T) \leq V(y, 2^{k+2}t)$, we then have

$$V(x, T) \leq D_0^{k+2} V(y, t) \leq D_0^2 (T/t)^{\nu_0} V(y, t).$$

Lemma 5.2.5 *Under the hypotheses of Theorem 5.2.3, there exists a constant C_1 such that, for any ball $B \subset M$,*

$$\forall\, f \in \mathcal{C}_0^\infty(B), \ \ \|f\|_p \leq C_1 r(B) \|\nabla f\|_p.$$

Fix a ball B and $s = 4r(B)$. Then, for all $f \in \mathcal{C}_0^\infty(B)$,

$$\left(\int_B |f(y) - f_s(y)|^p dy\right)^{1/p} \leq \|f - f_s\|_p \leq C_0 r(B) \|\nabla f\|_p. \qquad (5.2.2)$$

Observe that, for $y \in B$, $B(y, 4r(B)) \supset B$. Moreover, there exists a constant $\epsilon > 0$, independent of B and y, such that

$$V(y, 4r(B)) \geq (1 + \epsilon)\mu(B).$$

Indeed, let z be a point at distance $2r(B)$ from the center of B. Such a z exists because M is complete and not compact. Then $z \in B(y, 3r(B))$ and

$$B(z, r(B)) \subset B(y, 3r(B))$$

whereas $B(z, r(B)) \cap B = \emptyset$. By Lemma 5.2.4, it follows that

$$V(y, 4r(B)) \geq \mu(B) + V(z, r(B)) \geq (1 + \epsilon)\mu(B).$$

This shows that

$$
\begin{aligned}
f_s(y) &= \frac{1}{V(y, s)} \int_{B(y,s)} f(z) dz \\
&\leq \frac{1}{(1 + \epsilon)\mu(B)} \int_B f(z) dz \\
&\leq \frac{1}{(1 + \epsilon)} \left(\frac{1}{\mu(B)} \int_B |f(z)|^p dz\right)^{1/p} = \frac{\|f\|_p}{(1 + \epsilon)\mu(B)^{1/p}}.
\end{aligned}
$$

Now, write

$$\left(\int_B |f(y) - f_s(y)|^p dy\right)^{1/p} \geq \left(\int_B |f(y)|^p dy\right)^{1/p} - \left(\int_B |f_s(y)|^p dy\right)^{1/p}$$

$$\geq \|f\|_p - \frac{1}{(1+\epsilon)}\|f\|_p$$

$$= c\|f\|_p$$

where c depends only on ϵ. This and (5.2.2) show that

$$\|f\|_p \leq C_1 r(B)\|\nabla f\|_p$$

as desired.

With these two results at hand, the proof of Theorem 5.2.3 is very similar to that of Theorem 3.3.1. More precisely, fix a ball $B \subset M$. Observe that, by Lemma 5.2.4, for $y \in B$, $0 < s < r(B)$, and any $\nu \geq \nu_0$,

$$|f_s(y)| \leq \frac{1}{V(y,s)}\int_{B(y,s)} |f(z)|dz \leq \frac{C_0^2 r(B)^\nu}{\mu(B)s^\nu}\|f\|_1.$$

For any $C_0^\infty(B) \ni f \geq 0$ and any $t > 0$, write

$$\mu(f \geq \lambda) \leq \mu(\{|f - f_t| \geq \lambda/2\} \cap B) + \mu(\{f_t \geq \lambda/2\} \cap B)$$

and consider two cases.

Case 1: If λ is such that

$$\lambda > \frac{4D_0^2}{\mu(B)}\|f\|_1,$$

then pick $t < r(B)$ so that

$$\frac{D_0^2 r(B)^\nu}{\mu(B)t^\nu}\|f\|_1 = \lambda/4.$$

For this t,

$$\mu(\{f_t \geq \lambda/2\} \cap B) = 0$$

and

$$\mu(f \geq \lambda) \leq \mu(|f - f_t| \geq \lambda/2)$$
$$\leq (2/\lambda)^p\|f - f_t\|_p^p$$
$$\leq (2C_0 t\|\nabla f\|_2/\lambda)^p$$
$$= \left[C_2\mu(B)^{-1/\nu}r(B)\|f\|_1^{1/\nu}\|\nabla f\|_p \lambda^{-1-1/\nu}\right]^p.$$

That is

$$\lambda^{p(1+1/\nu)}\mu(f \geq \lambda) \leq \left[C_2\mu(B)^{-1/\nu}r(B)\|f\|_1^{1/\nu}\|\nabla f\|_p\right]^p.$$

Case 2: If λ is such that

$$\lambda \leq \frac{4D_0^2}{\mu(B)}\|f\|_1$$

then simply write (using Lemma 5.2.5)

$$\mu(f \geq \lambda) \leq \lambda^{-p}\|f\|_p^p \leq C_1\lambda^{-p}r(B)^p\|\nabla f\|_p^p.$$

It follows that

$$\lambda^{p(1+1/\nu)}\mu(f \geq \lambda) \leq \left[C_2'\mu(B)^{-1/\nu}r(B)\|f\|_1^{1/\nu}\|\nabla f\|_p\right]^p.$$

In both cases, we have

$$\lambda^{p(1+1/\nu)}\mu(f \geq \lambda) \leq \left[C_3\mu(B)^{-1/\nu}r(B)\|f\|_1^{1/\nu}\|\nabla f\|_p\right]^p$$

with $C_3 = \max\{C_2, C_2'\}$. Raising this inequality to the power τ/p with $\tau = 1/(1 + 1/\nu) = \nu/(1 + \nu)$ yields

$$\lambda\mu(f \geq \lambda)^{\tau/p} \leq (C_3\mu(B)^{-1/\nu}r(B))^\tau\|\nabla f\|_p^\tau\|f\|_1^{1-\tau}$$

which is $(S_{p/\tau,1}^{*\tau}(p))$. This, together with Theorem 3.2.2, proves Theorem 5.2.3 (see also the remark in Section 3.2.7). Note that one needs to take $\nu > p$ in order to be able to apply Theorem 3.2.2, i.e., in order to have the crucial parameter q in Theorem 3.2.2 be such that $0 < q < \infty$.

The same proof yields the following local result.

Theorem 5.2.6 *Assume that M is complete, not compact. Fix $R > 0$. Assume that M satisfies the pseudo-Poincaré inequality*

$$\forall f \in \mathcal{C}_0^\infty(M), \;\; \forall 0 < s < R, \;\; \|f - f_s\|_p \leq C_0 r\|\nabla f\|_p$$

where $f_s(x) = V(x,s)^{-1}\int_{B(x,s)} f d\mu$. Assume also that M satisfies the doubling volume growth condition

$$\forall x \in M, \;\; \forall 0 < t < R, \;\; V(x, 2t) \leq D_0 V(x, t).$$

Then there exist $\nu > 2$ and C such that

$$\forall f \in \mathcal{C}_0^\infty(B), \;\; \left(\int |f|^{p\nu/(\nu-p)}d\mu\right)^{(\nu-p)/\nu} \leq C\frac{r(B)^p}{\mu(B)^{p/\nu}}\int |\nabla f|^p d\mu$$

for all balls of radius at most $R/4$.

We end this section with a couple of remarks concerning doubling-type conditions.

Lemma 5.2.7 *Assume that there exist $R > 0$ and D_0 such that M satisfies the local doubling condition*

$$\forall\, x \in M, \ \ \forall\, r \in (0, R), \ \ \ V(x, 2r) \leq D_0 V(x, r).$$

Then there exists $C = C(D_0)$ such that for any $x, y \in M$ and $0 < r < R$,

$$V(x, r) \leq e^{Cd(x,y)/r} V(y, r).$$

Moreover, there exists $D_1 = D_1(D_0)$ such that

$$\forall\, x \in M, \ \ \forall\, T \geq R, \ \ \ V(x, T) \leq D_1^{T/R} V(x, R).$$

In particular, there exists $C = C(D_0)$ such that

$$\forall\, x \in M, \ \ \forall\, t, T \in (0, \infty), \ \ \ V(x, T) \leq (T/t)^{\nu_0} \exp(C(1 + T/R)) V(x, t)$$

where $\nu_0 = \log_2 D_0$.

Fix $0 < r < R$. Let B_1, B_2 be two balls of radius $r/4$ that intersect each other. Then, $B_1 \subset 4B_2$. Hence $\mu(B_1) \leq D_0^2 \mu(B_2)$. By symmetry, $\mu(B_2) \leq D_0^2 \mu(B_1)$. It easily follows that there exists $D_1 = D_1(D_0)$ such that any two balls B, B' with radii $s, s' \in [r/8, 2r]$ and at distance at most $10Nr$ from each other satisfy

$$D_1^{-N} \mu(B) \leq \mu(B') \leq D_1^N \mu(B). \tag{5.2.3}$$

In particular,

$$\mu(B(x, r)) \leq \exp\left(Cd(x, y)/r\right) \mu(B(y, r))$$

for all $x, y \in M$, $0 < r < R$. This proves the first assertion of Lemma 5.2.7.

To prove the second assertion it suffices to show that

$$\forall\, x \in M, \ \ \forall\, T \geq R, \ \ \ V(x, T + R/2) \leq D_1^{1/2} V(x, T). \tag{5.2.4}$$

Indeed, this will then give

$$\forall\, x \in M, \ \ \forall\, N = 1, 2, \ldots, \ \ \ V(x, (N+1)R) \leq D_1^N V(x, R).$$

Consider a maximal set X of points in $B(x, T - R/2)$ at distance at least $R/2$ apart. By definition, the balls $B(y, R/2)$, $y \in X$, are disjoint and all contained in $B(x, T)$. Thus,

$$\sum_{y \in X} V(y, R/2) \leq V(x, T). \tag{5.2.5}$$

As X is maximal, the balls $B(y, R)$, $y \in X$, cover $B(x, T - R/2)$. It follows that the balls $B(y, 2R)$, $y \in X$, cover $B(x, T + R/2)$. Thus

$$V(x, T + R/2) \leq \sum_{y \in X} V(y, 2R). \qquad (5.2.6)$$

By (5.2.5), (5.2.6) and the doubling hypothesis,

$$V(x, T + R/2) \leq D_0^2 V(x, T).$$

This proves (5.2.4). The last inequality in Lemma 5.2.7 then follows easily by Lemma 5.2.4.

The next lemma gives a lower bound on the volume growth of non-compact manifolds satisfying the doubling property.

Lemma 5.2.8 *Assume that M is a complete non-compact manifold which satisfies the doubling volume growth condition*

$$\forall x \in M, \quad \forall 0 < t < R, \quad V(x, 2t) \leq D_0 V(x, t).$$

Then there exist $\gamma > 0$ and C such that

$$\forall x \in M, \quad \forall 0 < t < s < R, \quad V(x, s) \leq C(s/t)^\gamma V(x, t).$$

By our previous lemma, we can assume that the doubling condition holds for all $0 < r < KR$ for some large fixed K. Now, consider some $r \in (0, KR)$ and the ball $B = B(x, r)$, $B' = B(x, r/2)$. Since M is connected and not compact, there exists a point y at distance $3r/4$ from x. Thus the ball $B(y, r/4)$ is contained in B and does not intersect B'. Moreover, by the doubling property, there exists $c > 0$ such that $\mu(B(y, r/4)) \geq c\mu(B')$. It follows that $\mu(B) \geq \mu(B') + \mu(B(y, r/4)) \geq (1 + c)\mu(B')$. The desired result easily follows by a dyadic iteration.

5.2.2 Mean value inequalities for subsolutions

The aim of this section is to show how localized Sobolev inequalities imply certain L^p mean value inequalities, $0 < p < \infty$, for subsolutions of the heat equation $(\partial_t + \Delta)u = 0$.

Fix a parameter $\tau > 0$. Consider $x \in M$, $r > 0$, $s \in \mathbb{R}$. Consider also a parameter δ, $0 < \delta < 1$ and set

$$
\begin{aligned}
Q(\tau, x, s, r) = Q &= (s - \tau r^2, s) \times B(x, r) \\
Q_\delta &= (s - \delta \tau r^2, s) \times B(x, \delta r).
\end{aligned}
$$

In what follows, we denote by $d\bar{\mu}$ the natural product measure on $\mathbb{R} \times M$:

$$d\bar{\mu} = dt \times d\mu.$$

Theorem 5.2.9 *Fix* $\tau > 0$. *Fix the ball* $B \subset M$ *of radius* $r = r(B) > 0$ *and center* x. *Assume that the local Sobolev inequality*

$$\left(\int |f|^{2\nu/(\nu-2)} d\mu \right)^{\nu/(\nu-2)} \leq C(B) \frac{r(B)^2}{\mu(B)^{2/\nu}} \int (|\nabla f|^2 + r(B)^{-2}|f|^2) d\mu$$

(5.2.7)

is satisfied for some $\nu > 2$ *and all* $f \in \mathcal{C}_0^\infty(B)$. *Fix* $0 < p < \infty$. *Then there exists a constant* $A(\tau, p, \nu)$ *such that, for any real* s, *any* $0 < \delta < \delta' \leq 1$, *and any smooth positive function* u *satisfying* $(\partial_t + \Delta)u \leq 0$ *in* $Q = Q(\tau, x, s, r)$, *we have*

$$\sup_{Q_\delta}\{u^p\} \leq \frac{A(\tau, p, \nu)C(B)^{\nu/2}}{(\delta' - \delta)^{2+\nu}r(B)^2\mu(B)} \int_{Q_{\delta'}} u^p d\bar{\mu}.$$

(5.2.8)

Before embarking on the proof, let us note that the hypothesis that u is smooth can be relaxed. For the proof given below, one only needs u to be locally bounded and in L^2 together with its first derivatives in time and space. In this case, the hypothesis $(\partial_t + \Delta)u \leq 0$ must be interpreted as meaning

$$\int [\partial_t u]\phi + \nabla u \cdot \nabla \phi \, d\bar{\mu} \leq 0$$

(5.2.9)

for all $\phi \geq 0$ in $L^2(Q)$ with $|\nabla \phi| \in L^2(Q)$ and such that $x \mapsto \phi(t, x)$ has compact support in B for all $t > 0$. By a purely local argument (in $\mathbb{R}^n$) any u which is locally in L^2 together with its first order space and time derivatives and satisfies (5.2.9) is indeed locally bounded. In fact, the local boundedness of u can be proved by an argument similar to the one developed below with a technical variation as for Theorem 2.2.2.

Another useful remark in applying the result above is that, for any solution u of $(\partial_t + \Delta)u = 0$ (possibly in the weak sense), the function $v = |u|$ is a non-negative subsolution, i.e., satisfies (5.2.9). To prove this without having ever to compute Δu, i.e., working with weak solutions, the best way is to show that $v_\epsilon = \sqrt{\epsilon + u^2}$ satisfies (5.2.9) and then let ϵ tend to 0. This is useful, for instance, in deriving bounds on $|\partial_t u|$ when u is a solution of $(\partial_t + \Delta)u = 0$. See Section 5.2.4 below.

We now proceed with the proof of Theorem 5.2.9. For simplicity, we assume for the proof that $\tau = \delta' = 1$. We first prove the case $p = 2$. The case $p > 2$ immediately follows since, for any smooth positive solution u of $(\partial_t + \Delta)u = 0$, $v = u^p$, $p \geq 1$ is also a smooth positive solution of $(\partial_t + \Delta)v \leq 0$. Indeed,

$$\Delta u^p = pu^{p-1}\Delta u - p(p-1)u^{p-2}|\nabla u|^2.$$

The case $0 < p < 2$ requires an additional argument as indicated below.

For any non-negative function $\phi \in \mathcal{C}_0^\infty(B)$, we have

$$\int_B [\phi \partial_t u + \nabla \phi \cdot \nabla u] \, d\mu \leq 0$$

(5.2.10)

which is just the integrated form of $(\partial_t + \Delta)u \leq 0$. For $\phi = \psi^2 u$, $\psi \in \mathcal{C}_0^\infty(B)$, we obtain

$$\int [\psi^2 u \partial_t u + \psi^2 |\nabla u|^2] d\mu \leq 2 \left| \int u \psi \nabla u \cdot \nabla \psi d\mu \right|$$

$$\leq 2 \int |\nabla \psi|^2 u^2 d\mu + \frac{1}{2} \int \psi^2 |\nabla u|^2 d\mu.$$

After some algebra, this yields

$$\int_B \left[2\psi^2 u \partial_t u + |\nabla(\psi u)|^2 \right] d\mu \leq A \|\nabla \psi\|_\infty^2 \int_{\mathrm{supp}(\psi)} u^2 d\mu \qquad (5.2.11)$$

where A is a numerical constant which will change from line to line in the argument developed below. For any smooth function χ of the time variable t, we easily get

$$\partial_t \left(\int_B [\chi \psi u]^2 d\mu \right) + \chi^2 \int_B |\nabla(\psi u)|^2 d\mu$$

$$\leq A\chi \left(\chi \|\nabla \psi\|_\infty^2 + \|\chi'\|_\infty \right) \int_{\mathrm{supp}(\psi)} u^2 d\mu.$$

We choose ψ and χ such that

$$0 \leq \psi \leq 1, \quad \mathrm{supp}(\psi) \subset \sigma B, \quad \psi = 1 \text{ in } \sigma'B, \quad |\nabla \psi| \leq 2(\tau r)^{-1},$$

$$0 \leq \chi \leq 1, \quad \chi = 0 \text{ in } (-\infty, s - \sigma r^2), \quad \chi = 1 \text{ in } (s - \sigma' r^2, \infty),$$

$$|\chi'| \leq 2(\tau r^2)^{-1},$$

where $0 < \sigma' < \sigma < 1$ and $\omega = \sigma - \sigma'$. Setting $I_\sigma = (s - \sigma r^2, s)$ and integrating our inequality over $(s - r^2, t)$ with $t \in I_{\sigma'}$, we obtain

$$\sup_{I_{\sigma'}} \left\{ \int_B \psi u^2 d\mu \right\} + \int \int_{I_{\sigma'} \times B} |\nabla(\psi u)|^2 d\bar\mu \leq A(r\omega)^{-2} \int \int_{Q_\sigma} u^2 d\bar\mu. \quad (5.2.12)$$

Let $E(B) = C(B)\mu(B)^{-2/\nu} r(B)^2$ be the Sobolev constant for the ball B given by (5.2.7) and set $q = \nu/(\nu - 2)$ where $\nu > 2$. Thanks to the Hölder inequality

$$\int w^{2(1+2/\nu)} \, d\mu \leq \left(\int w^{2q} \, d\mu \right)^{1/q} \left(\int w^2 \, d\mu \right)^{2/\nu},$$

(5.2.7) gives

$$\int w^{2(1+2/\nu)} \, d\mu \leq \left(\int w^2 \, d\mu \right)^{2/\nu} \left(E(B) \int \left[|\nabla w|^2 + r^{-2} |w|^2 \right] d\mu \right),$$

for all $w \in \mathcal{C}_0^\infty(B)$.

Returning to the subsolution u, the above inequality and (5.2.12) yield

$$\int\int_{Q_{\sigma'}} u^{2\theta} d\bar{\mu} \leq E(B) \left(A(r\omega)^{-2} \int\int_{Q_\sigma} u^2 d\bar{\mu} \right)^\theta \qquad (5.2.13)$$

with $\theta = 1 + 2/\nu$.

Now, for any $p \geq 1$, the function $v = u^p$ is also a smooth positive solution of $(\partial_t + \Delta)v \leq 0$. Therefore, (5.2.13) yields

$$\int\int_{Q_{\sigma'}} u^{2p\theta} d\bar{\mu} \leq E(B) \left(A(r\omega)^{-2} \int\int_{Q_\sigma} u^{2p} d\bar{\mu} \right)^\theta . \qquad (5.2.14)$$

Set $\omega_i = (1 - \delta)2^{-i}$ so that $\sum_1^\infty \omega_i = 1 - \delta$. Set also $\sigma_0 = 1$, $\sigma_{i+1} \doteq \sigma_i - \omega_i = 1 - \sum_1^i \omega_j$. Applying (5.2.14) with $p = p_i = \theta^i$, $\sigma = \sigma_i$, $\sigma' = \sigma_{i+1}$, we obtain

$$\int\int_{Q_{\sigma_{i+1}}} u^{2\theta^{i+1}} d\bar{\mu} \leq E(B) \left(A^{i+1}[(1 - \delta)r]^{-2} \int\int_{Q_{\sigma_i}} u^{2\theta^i} d\bar{\mu} \right)^\theta .$$

Hence,

$$\left(\int\int_{Q_{\sigma_{i+1}}} u^{2\theta^{i+1}} d\bar{\mu} \right)^{\theta^{-1-i}}$$

$$\leq A^{\sum(j+1)\theta^{-1-j}} E(B)^{\sum\theta^{-1-j}} [(1 - \delta)r]^{-2\sum\theta^{-j}} \int\int_Q u^2 d\bar{\mu}$$

where all the summations are taken from 1 to $i+1$. Letting i tend to infinity, we obtain

$$\sup_{Q_\delta}\{u^2\} \leq AE(B)^{\nu/2}[(1 - \delta)r]^{-2-\nu}\|u\|_{2,Q}^2. \qquad (5.2.15)$$

As $E(B) = C(B)\mu(B)^{-2/\nu}r(B)^2$, this yields (5.2.8) when $p \geq 2$.

The case $0 < p < 2$ follows from the case $p = 2$ by the argument used in the proof of Theorem 2.2.3 in Section 2.2.1. The only modification is that, in the present parabolic case, one must work with the cylinders Q_σ instead of the balls σB. This ends the proof of Theorem 5.2.9.

5.2.3 Localized heat kernel upper bounds

Consider the heat diffusion semigroup $H_t = e^{-t\Delta}$, $t > 0$, on M and its smooth positive heat kernel $h(t, x, y)$. We now show how Theorem 5.2.9 together with Lemma 4.2.1 yields certain Gaussian heat kernel upper bounds.

Theorem 5.2.10 *Let M be a complete non-compact manifold. There exists a constant A such that, for any $\epsilon \in (0,1)$ and any two balls $B_1 = B(x, r_1)$, $B_2 = B(y, r_2)$ satisfying (5.2.7) for some $\nu > 2$ with constant $C_1 = C(B_1)$ in B_1 (resp. $C_2 = C(B_2)$ in B_2), we have*

$$h(t, x, y) \leq \frac{AC_1C_2}{[\mu(B_1)\mu(B_2)]^{1/2}} \exp\left(-\frac{d(x,y)^2}{4t} + \epsilon \frac{d(x,y)}{\sqrt{t}} \right)$$

for all $t \geq \epsilon^{-2} \max\{r_1^2, r_2^2\}$.

Let $(H_t^{\alpha,\phi})_{t>0}$ be defined by

$$H_t^{\alpha,\phi} f(x) = e^{-\alpha\phi(x)} H_t[e^{\alpha\phi} f](x)$$

where $\phi \in \mathcal{C}_0^\infty(M)$ is a function satisfying $|\nabla \phi| \leq 1$ and α is a real parameter. See Section 4.2. By Lemma 4.2.1, we have

$$\|H_t^{\alpha,\phi}\|_{2 \to 2} \leq e^{\alpha^2 t}.$$

Fix $x, y \in M$ and $r_1, r_2 > 0$, and let χ_1 (resp. χ_2) be the function equal to 1 on $B_1 = B(x, r_1)$ (resp. $B_2 = B(y, r_2)$) and equal to 0 otherwise. Then

$$\int\!\!\int_{(\xi,\zeta) \in B_1 \times B_2} h(t, \xi, \zeta) e^{-\alpha(\phi(\xi) - \phi(\zeta))} d\xi d\zeta = \langle \chi_1, H_t^{\alpha,\phi} \chi_2 \rangle$$

$$\leq \|H_t^{\alpha,\phi}\|_{2 \to 2} \|\chi_1\|_2 \|\chi_2\|_2$$

$$\leq e^{\alpha^2 t} \mu(B_1)^{1/2} \mu(B_2)^{1/2}.$$

Using the fact that $|\phi| \leq 1$, we get

$$\int\!\!\int_{B_1 \times B_2} h(t, \xi, \zeta) d\xi d\zeta$$

$$\leq [\mu(B_1)\mu(B_2)]^{1/2} \exp(\alpha^2 t + \alpha(\phi(x) - \phi(y)) + |\alpha|(r_1 + r_2)).$$

As $u : (s, \zeta) \mapsto h(s, \xi, \zeta)$ is a positive solution of $(\partial_s + \Delta)u = 0$ in $(0, \infty) \times M$, assuming that $t \geq r_2^2$ and applying Theorem 5.2.9 with $p = 1$, we obtain

$$h(t, \xi, y) \leq \frac{AC_1}{r_2^2 \mu(B_2)} \int_{t-(1/4)r_2^2}^t \int_{B_2} h(s, \xi, \zeta) d\zeta ds.$$

Thus

$$\int_{B_1} h(t, \xi, y) d\xi \leq \frac{AC_1 \mu(B_1)^{1/2}}{\mu(B_2)^{1/2}} \exp(\alpha^2 t + \alpha(\phi(x) - \phi(y)) + |\alpha|(r_1 + r_2)).$$

By the same token, working with the variable ξ and assuming $t \geq r_1^2$, we get

$$h(t, x, y) \leq \frac{A^2 C_1 C_2}{[\mu(B_1)\mu(B_2)]^{1/2}} \exp(\alpha^2 t + \alpha(\phi(x) - \phi(y)) + |\alpha|(r_1 + r_2)).$$

Taking $\alpha = -(\phi(x) - \phi(y))/2t$ and assuming that $t \geq \epsilon^{-2} \max\{r_1^2, r_2^2\}$, we obtain

$$h(t, x, y) \leq \frac{A^2 C_1 C_2}{[\mu(B_1)\mu(B_2)]^{1/2}} \exp\left(-\frac{(\phi(x) - \phi(y))^2}{4t} + \epsilon\frac{|\phi(x) - \phi(y)|}{\sqrt{t}}\right).$$

Taking (as we may) a sequence of $\phi_i \in \mathcal{C}_0^\infty(M)$ with $|\nabla\phi_i| \leq 1$ and

$$\phi_i(x) - \phi_i(y) \to d(x, y)$$

finally gives

$$h(t, x, y) \leq \frac{A^2 C_1 C_2}{[\mu(B_1)\mu(B_2)]^{1/2}} \exp\left(-\frac{d(x, y)^2}{4t} + \epsilon\frac{d(x, y)}{\sqrt{t}}\right)$$

which is the desired result.

Corollary 5.2.11 *Let M be a complete non-compact manifold. Fix $R > 0$. Assume that there exist $\nu > 2$ and C such that, for any ball B of radius less than R, the local Sobolev inequality*

$$\left(\int |f|^{2\nu/(\nu-2)} d\mu\right)^{\nu/(\nu-2)} \leq C\frac{r(B)^2}{\mu(B)^{2/\nu}} \int (|\nabla f|^2 + r(B)^{-2}|f|^2)d\mu$$

is satisfied for all $f \in \mathcal{C}_0^\infty(B)$. Then there exists a constant A such that for all $x, y \in M$ and all $0 < t < R^2$,

$$h(t, x, y) \leq \frac{A}{[V(x, \frac{t}{\sqrt{t}+d(x,y)})V(y, \frac{t}{\sqrt{t}+d(x,y)})]^{1/2}} \exp\left(-\frac{d(x, y)^2}{4t}\right).$$

This follows from applying Theorem 5.2.10 with $B_1 = B(x, r_1)$, $B_2 = B(y, r_2)$, $r_1 = r_2 = \epsilon\sqrt{t}$, $\epsilon = (1 + d(x, y)/\sqrt{t})^{-1}$. Using Theorem 5.2.1, we can deduce from the bound above a slightly less precise but nicer looking estimate, namely, for all $x, y \in M$, $0 < t < R^2$,

$$h(t, x, y) \leq \frac{A(1 + d(x, y)/\sqrt{t})^\nu}{[V(x, \sqrt{t})V(y, \sqrt{t})]^{1/2}} \exp\left(-\frac{d(x, y)^2}{4t}\right). \tag{5.2.16}$$

Using Lemma 5.2.7, one also obtains that for any $\epsilon > 0$ there exists A_ϵ such that

$$h(t, x, y) \leq \frac{A_\epsilon}{V(x, \sqrt{t})} \exp\left(-\frac{d(x, y)^2}{4(1 + \epsilon)t}\right) \tag{5.2.17}$$

for all $x, y \in M$, $0 < t < R^2$.

Corollary 5.2.12 *Let M be a complete non-compact manifold. Assume that there exist $\nu > 2$ and C such that (5.2.1) holds true for $p = 2$ with $\sup_B C(B) = C < \infty$. Then there exists a constant A such that for all $x, y \in M$ and all $0 < t < \infty$,*

$$h(t, x, y) \leq \frac{A}{[V(x, \frac{t}{\sqrt{t+d(x,y)}})V(y, \frac{t}{\sqrt{t+d(x,y)}})]^{1/2}} \exp\left(-\frac{d(x,y)^2}{4t}\right).$$

Moreover, one also has

$$h(t, x, y) \leq \frac{A(1 + d(x,y)/\sqrt{t})^{2\nu}}{V(x, \sqrt{t})} \exp\left(-\frac{d(x,y)^2}{4t}\right).$$

The first bound follows readily from Corollary 5.2.11. To obtain the second bound, we use the first bound and the fact that

$$V(y, r) \geq c\left(\frac{r}{d(x,y)+r}\right)^\nu V(y, d(x,y)+r) \geq c\left(\frac{r}{r+d(x,y)}\right)^\nu V(x, r)$$

for all $x, y \in M$, $r > 0$. See Theorem 5.2.2.

Such precise Gaussian upper bounds lead to an optimal on-diagonal lower bound for $h(t, x, x)$ as explained in Section 4.2.4. For this, one needs the next elementary lemma.

Lemma 5.2.13 *Assume that M is complete, not compact, and satisfies the doubling volume growth condition*

$$\forall x \in M, \quad \forall t \in (0, R), \quad V(x, 2t) \leq D_0 V(x, t).$$

Then there exist $C, c > 0$ such that, for all $x \in M$, $t \in (0, R^2)$ and $r \in (0, R)$,

$$cV(x, \sqrt{t})e^{-3r^2/t} \leq \int_{d(x,y)\geq r} e^{-d(x,y)^2/t} dy \leq CV(x, \sqrt{t})e^{-r^2/(2t)}.$$

For the proof, fix $x \in M$, $t \in (0, R^2)$, $r \in (0, R)$, and set

$$I(x, t) = \int_{d(x,y)\geq r} e^{-d(x,y)^2/t} dy.$$

For the lower bound, consider a point z at distance $2r + \sqrt{t}$ from x. Then

$$
\begin{aligned}
I(x, t) &\geq \int_{B(z, r+\sqrt{t})} e^{-d(x,y)^2/t} dy \\
&\geq c_1 V(z, r + \sqrt{t})e^{-3r^2/t} \\
&> c_2 V(z, 2(r + \sqrt{t}))e^{-3r^2/t} \\
&\geq c_3 V(x, \sqrt{t})e^{-3r^2/t}.
\end{aligned}
$$

For the upper bound, observe that it suffices to consider the case when $\sqrt{t} < r$ because

$$\int_{B(x,\sqrt{t})} e^{-d(x,y)^2/t} dy \leq V(x,\sqrt{t}).$$

Assuming $\sqrt{t} < r$, write

$$
\begin{aligned}
I(x,t) &\leq \sum_{k=1}^{\infty} \int_{r2^{k-1} < d(x,y) \leq r2^k} \exp\left(-\frac{d(x,y)^2}{t}\right) dy \\
&\leq \sum_{k=1}^{\infty} V(x,r2^k) \exp\left(-\frac{r^2 2^{2(k-1)}}{t}\right) \\
&\leq V(x,\sqrt{t}) \sum_{k=1}^{\infty} \frac{V(x,r2^k)}{V(x,\sqrt{t})} \exp\left(-\frac{r^2 2^{2(k-1)}}{t}\right) \\
&\leq V(x,\sqrt{t}) \sum_{k=1}^{\infty} \exp\left(C\frac{r2^k}{\sqrt{t}} - \frac{r^2 2^{2(k-1)}}{t}\right).
\end{aligned}
$$

The last inequality follows from Lemma 5.2.7 which yields, under the volume hypothesis of Lemma 5.2.13,

$$\forall x \in M, \ \forall s \in (0,R), \ \forall T \in (s,\infty), \quad \frac{V(x,T)}{V(x,s)} \leq \exp\left(CT/s\right).$$

Since

$$\sum_{k=1}^{\infty} \exp\left(C\frac{r2^k}{\sqrt{t}} - \frac{r^2 2^{2(k-1)}}{t}\right) \leq \exp\left(-C'\frac{r^2}{\sqrt{t}}\right),$$

the desired result follows.

With the help of Lemma 5.2.13, the argument of Section 4.2.4 can be adapted to prove the following theorem.

Theorem 5.2.14 *Let M be a complete non-compact manifold. Fix $R \in (0,\infty]$. Assume that there exist $\nu > 2$ and C such that, for any ball B of radius less than R, the local Sobolev inequality*

$$\left(\int |f|^{2\nu/(\nu-2)} d\mu\right)^{\nu/(\nu-2)} \leq C \frac{r(B)^2}{\mu(B)^{2/\nu}} \int (|\nabla f|^2 + r(B)^{-2}|f|^2) d\mu$$

is satisfied for all $f \in \mathcal{C}_0^{\infty}(B)$. Then there exist two constants $c, \epsilon > 0$ such that for all $x \in M$ and all $0 < t < \epsilon R^2$,

$$h(t,x,x) \geq \frac{c}{V(x,\sqrt{t})}.$$

In particular, if there exist $\nu > 2$ and C such that (5.2.1) holds true for $p = 2$ with $\sup_B C(B) = C < \infty$, then the lower bound

$$h(t,x,x) \geq \frac{c}{V(x,\sqrt{t})}$$

holds true for all $x \in M$ and $t > 0$.

We refer the interested reader to [18] for a thorough discussion of on-diagonal heat kernel lower bounds. Here we simply note that the hypotheses of the above theorem are not sufficient to imply a Gaussian lower bound of the type

$$h(t, x, y) \geq \frac{c}{V(x, \sqrt{t})} \exp\left(-\frac{Cd(x, y)^2}{t}\right). \tag{5.2.18}$$

A counterexample is obtained by gluing two copies of $\mathbb{R}^3$ through a small compact cylinder. Let o be a fixed point near this compact cylinder. It is not hard to see that the manifold M obtained in this way satisfies the usual 3-dimensional Sobolev inequality and has volume growth $V(x, r) \approx r^3$ for all $r > 0$. Thus it satisfies the hypotheses of Corollary 5.2.12 and Theorem 5.2.14. It can also be shown that the Gaussian lower bound (5.2.18) fails in this case when t, x and y are such that $d(o, x) \approx d(o, y) \approx \sqrt{t} \to \infty$ with x, y each in one of the two different copies of $\mathbb{R}^3$. In this case, $h(t, x, y)$ is actually of order t^{-2} instead of $t^{-3/2}$. See [35]. Heuristically, a Brownian particle trying to go from x to y is much less likely to succeed in M than, say, in $\mathbb{R}^3$. This is because, in M, the particle has to go through a fixed compact neighborhood of the central point o in order to pass from one copy of $\mathbb{R}^3$ to the other.

5.2.4 Time-derivative upper bounds

The results obtained so far easily yield some time-derivative estimates for positive solutions of the heat equation. This can be seen as follows. Let u be a non-negative subsolution of $(\partial_t + \Delta)u = 0$ in a cylinder $Q = (t - r^2, s) \times B(x, r)$. Set $B = B(x, r)$. From (5.2.11), one easily extracts

$$\int_{\delta B} |\partial_t u|^2 d\mu \leq A(1 - \delta)^{-2} r^{-2} \int_B |u|^2 d\mu.$$

If u is a positive solution of $(\partial_t + \Delta)u = 0$ in Q, then $\partial_t^k u$ is a solution of the same equation in Q and $v = |\partial_t^k u|$ is a non-negative subsolution. Moreover, $v|\partial_t v| = v|\partial_t^{k+1} u|$. Indeed, $|\partial_t v|$ and $|\partial_t^{k+1} u|$ differ only when $\partial^k u = 0$. Thus we can again apply (5.2.11) to obtain

$$\int_{\delta B} |\partial_t^{k+1} u|^2 d\mu \leq A(1 - \delta)^{-2} r^{-2} \int_B |\partial_t^k u|^2 d\mu.$$

It follows that, as long as $(1 - \delta)k = 1 - \sigma$ with $0 < \sigma < 1$,

$$\int_{\sigma B} |\partial_t^k u|^2 d\mu \leq A k^{2k} (1 - \sigma)^{-2k} r^{-2k} \int_B |u|^2 d\mu.$$

(Note that in the argument above one works with subsolutions that are not smooth; see the remark after Theorem 5.2.9.)

Assuming that (5.2.7) is satisfied and using Theorem 5.2.9 for the sub-solution $|\partial_t^k u|$, we conclude that

$$\sup_{Q_\sigma}\{|\partial_t^k u|^2\} \leq A(\sigma,\nu,k)C(B)^{\nu/2}r^{-2k-1}\mu(B)^{-1}\int_Q |u|^2 d\bar\mu.$$

Here, $Q = (t - r^2, s) \times B(x,r)$ and $Q_\sigma = (t - \sigma r^2, s) \times B(x, \sigma r)$.

Let us apply this to the heat kernel $h(t, x, y)$ in the case where we have a full scale of localized Sobolev inequalities as in Corollary 5.2.12. Using the argument above and the heat kernel bound given by Corollary 5.2.12, we obtain the following result.

Theorem 5.2.15 *Let M be a complete non-compact manifold. Assume that there exist $\nu > 2$ and C such that (5.2.1) holds true for $p = 2$ with $\sup_B C(B) = C < \infty$. Then, for each $k = 0, 1, 2, \ldots$, and $\epsilon > 0$, there exists a constant $A = A(k, \epsilon)$ such that for all $x, y \in M$ and all $0 < t < \infty$,*

$$|\partial_t^k h(t, x, y)| \leq \frac{A}{t^k V(x, \sqrt{t})}\exp\left(-\frac{d(x,y)^2}{4t(1+\epsilon)}\right).$$

If the hypothesis holds only for balls of radius $r < R$ then the conclusion above holds for all $x, y \in M$ and all $0 < t < R^2$.

This method is by no means the only way to derive upper bounds for the time-derivatives of the heat kernel. See [22] for a powerful and widely applicable technique.

5.2.5   Mean value inequalities for supersolutions

The main tool used in the preceding section is the mean value inequality for subsolutions stated in Theorem 5.2.9. Supersolutions satisfy similar but different inequalities that are presented below. In the statement below, we assume for simplicity that u is smooth. This hypothesis (which is very unnatural for supersolutions) can be relaxed to the requirement that u is locally in L^2 together with its space and time first derivatives without essential change in the proof (u is then a supersolution in an obvious weak sense and one has to perform integration in the time variable sooner in the argument than we will do below, but this poses no difficulty). Let us note here that, unlike subsolutions, supersolutions need not be locally bounded.

Fix a parameter $\tau > 0$. Consider $x \in M$, $r, s > 0$ and a small positive parameter $0 < \delta < 1$ and recall the notation

$$\begin{aligned} Q(\tau, x, s, r) = Q &= (s - \tau r^2, s) \times B(x, r), \\ Q_\delta &= (s - \delta \tau r^2, s) \times B(x, \delta r), \\ d\bar\mu &= dt \times d\mu. \end{aligned}$$

Theorem 5.2.16 *Fix $\tau > 0$. Fix the ball $B \subset M$ of radius $r = r(B) > 0$ and center x. Assume that the local Sobolev inequality (5.2.7) is satisfied for some $\nu > 2$. Then there exists a constant $A(\tau, \nu)$ such that, for any real s, any $0 < \delta < \delta' \leq 1$, any $0 < p < \infty$, and any smooth positive function u satisfying $(\partial_t + \Delta)u \geq 0$ in $Q = Q(\tau, x, s, r)$, we have*

$$\sup_{Q_\delta}\{u^{-p}\} \leq \frac{A(\tau, \nu)C(B)^{\nu/2}}{(\delta' - \delta)^{2+\nu}r(B)^2\mu(B)} \int_{Q_{\delta'}} u^{-p}d\bar{\mu}. \tag{5.2.19}$$

For simplicity, we assume for the proof that $\tau = \delta' = 1$. For any non-negative function $\phi \in \mathcal{C}_0^\infty(B)$, we have

$$\int_B [\phi\partial_t u + \nabla\phi \cdot \nabla u] \, d\mu \geq 0. \tag{5.2.20}$$

For $\phi = p\psi^2 u^{-p-1}$, $\psi \in \mathcal{C}_0^\infty(B)$, $w = u^{-p/2}$ this yields

$$-\int [\psi^2\partial_t w^2 + 4\frac{p+1}{p}\psi^2|\nabla w|^2 + 4w\psi\nabla\psi \cdot \nabla w]d\mu \geq 0.$$

As $1 < (p+1)/p < \infty$, elementary algebra and the inequality $|ab| \leq (1/2)(a^2 + b^2)$ yield

$$\int_B [\psi^2\partial_t w^2 + |\nabla(\psi w)|^2] \, d\mu \leq A\|\nabla\psi\|_\infty^2 \int_{\mathrm{supp}(\psi)} w^2 d\mu \tag{5.2.21}$$

where A is a numerical constant which will change from line to line. For any $0 < \sigma' < \sigma < 1$, the argument used to obtain (5.2.13) applies here and yields

$$\int\int_{Q_{\sigma'}} w^{2\theta}d\bar{\mu} \leq E(B)\left(A(r\omega)^{-2}\int\int_{Q_\sigma} w^2 \, d\bar{\mu}\right)^\theta$$

with $\theta = 1 + 2/\nu$, $\omega = \sigma - \sigma'$ and $E(B) = C(B)\mu(B)^{-2/\nu}r(B)^2$.

In terms of the supersolution u, this reads

$$\int\int_{Q_{\sigma'}} u^{-p\theta}d\bar{\mu} \leq E(B)\left(A(r\omega)^{-2}\int\int_{Q_\sigma} u^{-p} \, d\bar{\mu}\right)^\theta \tag{5.2.22}$$

for any $0 < p < \infty$. From here, the iteration used to prove (5.2.15) yields

$$\sup_{Q_\delta}\{u^{-p}\} \leq AE(B)^{\nu/2}[(1-\delta)r]^{-2-\nu}\int\int_Q u^{-p}d\bar{\mu} \tag{5.2.23}$$

which is the desired result.

In order to state the next result we need to introduce the following notation. Given $x \in M$ and reals τ, s, r, δ with $\tau, r > 0$ and $0 < \delta < 1$, we set

$$Q'_\delta(\tau, x, s, r) = (s - \tau r^2, s - (1-\delta)\tau r^2) \times \delta B.$$

Theorem 5.2.17 *Fix the ball $B \subset M$ of radius $r = r(B) > 0$ and center x. Assume that the local Sobolev inequality (5.2.7) is satisfied for some $\nu > 2$ and set $\theta = 1 + 2/\nu$. Fix $\tau > 0$. Fix also $0 < p_0 < \theta$. Then there exists a constant $A(p_0, \tau, \nu)$ such that, for any real s, any $0 < \delta < \delta' < 1$, any $0 < p < p_0/\theta$, and any smooth positive function u satisfying $(\partial_t + \Delta)u \geq 0$ in $Q = Q(\tau, x, s, r)$, we have*

$$\left(\int_{Q'_\delta} u^{p_0} d\bar{\mu} \right)^{p/p_0} \leq \left[\frac{A(p_0, \tau, \nu)C(B)^{1+\nu}}{(\delta' - \delta)^{4+2\nu} r(B)^2 \mu(B)} \right]^{1-p/p_0} \int_{Q'_{\delta'}} u^p d\bar{\mu}. \quad (5.2.24)$$

We give the proof assuming $\delta' = \tau = 1$. In (5.2.20), we set $\phi = \alpha \psi^2 u^{\alpha+1}$ with $\psi \in \mathcal{C}_0^\infty(B)$ and $0 < \alpha < p_0(1 + 2/\nu)^{-1}$. We also set $w = u^{\alpha/2}$. This yields

$$\int [\psi^2 \partial_t w^2 + 4\frac{\alpha - 1}{\alpha} \psi^2 |\nabla w|^2 + 4w\psi \nabla \psi \cdot \nabla w] d\mu \geq 0.$$

Note that $\alpha - 1$ is negative and that

$$\frac{|\alpha - 1|}{\alpha} \geq 1 - p_0/\theta = \epsilon > 0.$$

This easily yields

$$-\int_B \left[\psi^2 \partial_t w^2 + |\nabla(\psi w)|^2 \right] d\mu \leq A_\epsilon \|\nabla \psi\|_\infty^2 \int_{\text{supp}(\psi)} w^2 d\mu.$$

This should be compared with (5.2.21). The difference with (5.2.21) is the minus sign appearing in front of the first integral above. This difference explains why we have to work with the cylinders Q'_δ (which are chopped at the top) instead of Q_δ (which are chopped at the bottom). Apart from this, the same argument used to obtain (5.2.13) applies here again and yields

$$\int \int_{Q'_{\sigma'}} u^{\alpha\theta} d\bar{\mu} \leq E(B) \left(A(r\omega)^{-2} \int \int_{Q'_\sigma} u^\alpha d\bar{\mu} \right)^\theta \quad (5.2.25)$$

for all $0 < \alpha < p_0(1 + 2/\nu)^{-1}$, $0 < \sigma' < \sigma < 1$, with $\omega = \sigma - \sigma'$ and $E(B) = C(B)\mu(B)^{-2/\nu} r(B)^2$. Compare with (5.2.22). To finish the proof, it now suffices to repeat the iteration argument appearing after (2.2.8).

5.3   Poincaré inequalities

In Section 5.2, applications of local Sobolev inequalities, including heat kernel upper bounds, were developed. We observed there that Gaussian heat kernel lower bounds cannot be obtained from such Sobolev inequalities alone. The crucial missing tool for obtaining Gaussian heat kernel lower bounds is Poincaré inequality.

We say that a complete manifold M satisfies a (scale-invariant) Poincaré inequality if there exist two constants P_0 and $\kappa \geq 1$ such that, for any ball $B \subset M$ of radius $r(B) > 0$,

$$\forall f \in \mathcal{C}^\infty(B), \quad \int_B |f - f_B|^2 d\mu \leq P_0 r(B)^2 \int_{\kappa B} |\nabla f|^2 d\mu$$

where f_B is the mean of f over B.

It is useful to generalize this definition and introduce two extra parameters $1 \leq p < \infty$ and $R > 0$. We say that a complete manifold M satisfies a scale-invariant L^p Poincaré inequality up to radius R if there exist two constants P_0 and $\kappa \geq 1$ such that, for any ball $B \subset M$ of radius $0 < r(B) \leq R$,

$$\forall f \in \mathcal{C}^\infty(B), \quad \int_B |f - f_B|^p d\mu \leq P_0 r(B)^p \int_{\kappa B} |\nabla f|^p d\mu$$

where f_B is the mean of f over B.

One crucial aspect of these inequalities is that they are assumed to hold for all smooth $f \in \mathcal{C}^\infty(B)$ instead of merely $f \in \mathcal{C}_0^\infty(B)$.

5.3.1   Poincaré inequality and Sobolev inequality

The main result of this section is Theorem 5.3.3, which shows that Poincaré inequality and the doubling property of the measure imply a family of local Sobolev inequalities. This is one of the key technical points needed to apply Moser's iterative technique under the assumption that Poincaré inequality and doubling are satisfied.

We start with an easy lemma.

Lemma 5.3.1 *Fix $1 \leq p < \infty$ and $0 < R \leq \infty$. Assume that there exist two constants P_0 and $\kappa \geq 1$ such that, for any ball $B \subset M$ of radius $0 < r(B) \leq R$,*

$$\forall f \in \mathcal{C}^\infty(M), \quad \int_B |f - f_B|^p d\mu \leq P_0 r(B)^p \int_{\kappa B} |\nabla f|^p d\mu \qquad (5.3.1)$$

where f_B is the mean of f over B. Assume also that M satisfies the doubling condition

$$\forall x \in M, \ \forall 0 < t < R, \quad V(x, 2t) \leq D_0 V(x, t). \qquad (5.3.2)$$

Then, for any $1 < \tau \leq \kappa$ and any $K > 1$ there exists a constant C depending on P_0, κ, D_0, τ and K such that

$$\forall f \in \mathcal{C}^\infty(M), \quad \int_B |f - f_B|^p d\mu \leq C r(B)^p \int_{\tau B} |\nabla f|^p d\mu$$

for any ball B of radius less than KR.

We only give the outline of the proof and leave the details to the reader. Fix $x \in M$ and $0 < r < KR$ and $0 < \tau < \kappa$. By a well-known argument, one can cover the ball $B = B(x, r)$ by a finite collection of balls of radius $\theta = \min\{(\tau - 1)r/(100\kappa), R\}$ with center in B and such that, for any two balls A, A' in this collection, the balls $\frac{1}{2}A, \frac{1}{2}A'$ are disjoint. Moreover, by (5.3.2) and Lemma 5.2.7, the cardinal of such a collection of balls is uniformly bounded, independently of B. From this and a chaining argument, the desired result follows. The chaining argument alluded to here is a simpler version of what is done below in (5.3.10), (5.3.11). Of course, one of the points of this argument is that the balls $100\kappa B$ are all contained in τB.

Lemma 5.3.2 *Fix $1 \leq p < \infty$ and $0 < R \leq \infty$. Assume that M satisfies a scale-invariant L^p Poincaré inequality up to radius R, i.e., assume that (5.3.1) is satisfied. Assume further that M satisfies the doubling condition (5.3.2). Then*

$$\|f - f_s\|_p \leq Cs\|\nabla f\|_p$$

for all $f \in \mathcal{C}_0^\infty(M)$ and all $0 < s < R/4$. That is, M satisfies an L^p pseudo-Poincaré inequality.

By Lemma 5.3.1, we can assume that $\kappa = 2$. Fix $0 < s < R/4$. Let B_j be a collection of balls of radius $s/2$ such that $B_i \cap B_j = \emptyset$ and $M = \bigcup_i 2B_i$. Such a collection always exists (using Zorn's lemma). The doubling condition implies easily that the overlapping number

$$N(z) = \#\{i : z \in 8B_i\}$$

satisfies

$$\sup_{z \in M} N(z) = N_0 < \infty.$$

Now, write

$$
\begin{aligned}
\|f - f_s\|_p^p &\leq \sum_i \int_{2B_i} |f(x) - f_s(x)|^p d\mu \\
&\leq 2^p \sum_i \int_{2B_i} \left(|f(x) - f_{4B_i}|^p + |f_s(x) - f_{4B_i}|^p\right) d\mu. \quad (5.3.3)
\end{aligned}
$$

By the postulated Poincaré inequality, we have

$$
\begin{aligned}
\sum_i \int_{2B_i} |f(x) - f_{4B_i}|^p d\mu &\leq \sum_i \int_{4B_i} |f(x) - f_{4B_i}|^p d\mu \\
&\leq C_1 s^p \sum_i \int_{8B_i} |\nabla f|^p d\mu \\
&\leq C_1 N_0 s^p \int_M |\nabla f|^p d\mu. \quad (5.3.4)
\end{aligned}
$$

By the doubling condition and Poincaré inequality, we also have

$$
\begin{aligned}
\sum_i \int_{2B_i} |f_s(x) - f_{4B_i}|^p d\mu &\leq \sum_i \int_{2B_i} \frac{1}{V(x,s)} \left[\int_{B(x,s)} |f(y) - f_{4B_i}|^p dy \right] dx \\
&\leq C_2 \sum_i \frac{1}{\mu(B_i)} \int_{2B_i} \int_{4B_i} |f - f_{4B_i}|^p d\mu d\mu \\
&\leq C_3 s^p \sum_i \int_{8B_i} |\nabla f|^p d\mu \\
&\leq C_3 N_0 \, s^p \int_M |\nabla f|^p d\mu. \qquad (5.3.5)
\end{aligned}
$$

By (5.3.3), (5.3.4) and (5.3.5), the desired inequality

$$
\|f - f_s\|_p \leq Cs \|\nabla f\|_p
$$

follows.

We can now state the main result of this section.

Theorem 5.3.3 *Fix $1 \leq p < \infty$ and $0 < R \leq \infty$. Assume that M satisfies a scale-invariant L^p Poincaré inequality up to radius R, i.e., assume that (5.3.1) is satisfied. Assume further that M satisfies the doubling condition (5.3.2). Then, for any $K > 1$, there exist $\nu > p$ and C such that, for any ball B of radius less than KR,*

$$
\forall f \in \mathcal{C}_0^\infty(B), \quad \left(\int |f|^{p\nu/(\nu-p)} d\mu \right)^{(\nu-p)/\nu} \leq C \frac{r(B)^p}{\mu(B)^{p/\nu}} \int |\nabla f|^p d\mu. \quad (5.3.6)
$$

This follows from Lemma 5.3.2, Theorem 5.2.6 and Lemmas 5.2.7, 5.3.1.

5.3.2 Some weighted Poincaré inequalities

The results contained in this section are important technical tools. We show that, if M has the doubling property (5.3.2) and satisfies the Poincaré inequality (5.3.1), then one can always take $\kappa = 1$ in (5.3.1). This fact is due to D. Jerison [46]. The idea is to use a Whitney-type covering of the ball B. Jerison's proof uses a rather subtle analysis of the covering near the boundary. It was later observed by Guozhen Lu [57] that a simpler argument can be given based on ideas from earlier works of Bojarski [8] and Chua [16] on Euclidean domains. This argument has been used by many authors in various settings, e.g., [25].

We also produce some useful weighted Poincaré inequalities, for which we need the following notation. The weights we are interested in are rather simple. Their introduction appears to be crucial in the last part of Moser's iteration argument for parabolic equations. For $R > 0$, $\alpha \in (0,1)$, let $\mathcal{M}(R, \alpha)$ be the set of all non-increasing functions $\phi : [0, \infty) \to [0, 1]$ such that:

- $\inf\{s > 0 : \phi(s) = 0\} = R$

- $\forall 0 < s < R, \quad \phi(s + (1/2)[(R - s) \wedge (R/2)]) \geq \alpha\phi(s).$

Two interesting examples of such functions ϕ are:

- $\phi(s) = 1$ on $[0, R]$ and 0 otherwise, which belongs to $\mathcal{M}(R, 1)$.

- $\phi(s) = (1 - s/R)_+^\gamma$ for some $\gamma > 0$, which belongs to $\mathcal{M}(R, (1/4)^\gamma)$.

Thus, functions in $\mathcal{M}(R, \alpha)$ may vanish at $s = R$ but, if they do, they do so in such a way that $(R - s) \approx (R - t)$ implies $\phi(s) \approx \phi(t)$.

Given a function $\phi \in \mathcal{M}(R, \alpha)$ and $x \in M$, we set

$$\Phi(y) = \Phi_x(y) = \phi(d(x, y)).$$

Theorem 5.3.4 *Fix* $1 \leq p < \infty$, $\alpha \in (0, 1)$ *and* $R \in (0, \infty]$. *Assume that* M *satisfies a scale-invariant* L^p *Poincaré inequality up to radius* R, *i.e., assume that* (5.3.1) *is satisfied. Assume further that* M *satisfies the doubling condition* (5.3.2). *Then there exists a constant* P_α *such that, for all* $x \in M$, *all* $0 < r < R$, *and all functions* $\phi \in \mathcal{M}(r, \alpha)$ *with* $0 < r < R$, *we have*

$$\forall f \in \mathcal{C}^\infty(M), \quad \int |f - f_\Phi|^p \Phi d\mu \leq P_\alpha r^p \int |\nabla f|^p \Phi d\mu$$

where $f_\Phi = \int f \Phi d\mu / \int \Phi d\mu$ *and* $\Phi(y) = \phi(d(x, y))$.

Corollary 5.3.5 *Fix* $1 \leq p < \infty$ *and* $R > 0$. *Assume that* (5.3.1) *and* (5.3.2) *are satisfied. Then there exists a constant* P *such that, for all* $x \in M$ *and all* $0 < r < R$ *we have*

$$\forall f \in \mathcal{C}^\infty(M), \quad \int_B |f - f_B|^p d\mu \leq P r^p \int_B |\nabla f|^p d\mu$$

where f_B *is the mean of* f *over* B.

The proof of these results will be given below. The main ingredient is a somewhat subtle covering argument that will be expained in detail. Here, let us observe that Lemma 5.3.1 shows that we can always decrease the constant κ appearing in (5.3.1) to any specified value strictly larger than 1 at the expense of a larger but still finite P_0. What is not obvious but is achieved in Theorem 5.3.4 and Corollary 5.3.5 is that one can in fact take $\kappa = 1$. We will use Lemma 5.3.1 to simplify the proof of Theorem 5.3.4 by assuming that (5.3.1) holds with $\kappa = 2$.

5.3.3   Whitney-type coverings

All the balls considered below are open balls, i.e.,

$$B(x,r) = \{z \in M : d(x,z) < r\}.$$

We will use without further comment the fact that for any two points x, y there exists a distance-minimizing curve joining x to y in M. In this subsection we assume that the doubling condition (5.3.2) is satisfied for some fixed $R > 0$.

Let us fix a ball $E = B(x,r)$, $x \in M$, $0 < r < R$. We claim that there exists a collection $\mathcal{F}$ of balls B having the following properties:

(1) The balls $B \in \mathcal{F}$ are disjoint.

(2) The balls $2B$, $B \in \mathcal{F}$, form a covering of E, i.e., $E = \bigcup_{B \in \mathcal{F}} 2B$.

(3) For any ball $B \in \mathcal{F}$, $r(B) = 10^{-3} d(B, \partial E)$. In particular, $10^3 B \subset E$.

(4) There exists a constant K depending only on the constant D_0 in (5.3.2) such that

$$\sup_{\eta \in E} \#\{B \in \mathcal{F} : \eta \in 10^2 B\} \leq K. \qquad (5.3.7)$$

We will call $\mathcal{F}$ a covering of E (although only the balls $2B$, $B \in \mathcal{F}$ actually cover E).

To construct $\mathcal{F}$, start with the collection $\overline{\mathcal{F}}$ of all balls B with center in E and radius $r(B) = 10^{-3} d(B, \partial E)$. Let us start by noting that for each $z \in E$ there exists a ball $B \in \overline{\mathcal{F}}$ with center z. Indeed, for any B with center z, the condition

$$d(B, \partial E) = 10^3 r(B)$$

is the same as

$$d(z, \partial E) = (1 + 10^3) r(B).$$

Start $\mathcal{F}$ by picking a ball B_0 in $\overline{\mathcal{F}}$ with the largest possible radius. Such a ball exists by a simple compactness argument; see below. Then pick the next ball B_1 in $\mathcal{F}$ to be a ball in $\overline{\mathcal{F}}$ which does not intersect B_0 and has maximal radius. Assuming that k balls $B_0, B_1, \ldots, B_{k-1}$ have already been picked, pick the next ball B_k to be a ball in $\overline{\mathcal{F}}$ which does not intersect $\bigcup_0^{k-1} B_i$ and has maximal radius.

Let us show that such a ball does exist. Let $B_j = B(x_j, r_j)$, $0 \leq j \leq k-1$. Let ρ be the least upper bound of the radii of balls in $\overline{\mathcal{F}}$ that do not intersect $W_{k-1} = \bigcup_0^{k-1} B_i$. Then, there exist two sequences $z_j \in M$, $\rho_j > 0$ and a point $z \in E$ such that $B(z_j, \rho_j) \in \overline{\mathcal{F}}$, $B(z_j, \rho_j) \cap W_{k-1} = \emptyset$, $z_j \to z$, $\rho_j \to \rho$. Consider the ball $B_k = B(z, \rho)$. By continuity, $d(x_k, z) \geq r_k$. Thus $B(z, \rho)$ does not intersect $\bigcup_0^{k-1} B_i$. For each j, let y_j be a point such that $d(z_j, y_j) = \rho_j$ and $d(B(z_j, \rho_j), \partial E) = d(y_j, \partial E)$. By extracting a subsequence, we can assume that $y_j \to y$. Then, by continuity $d(z, y) = \rho$ and $d(B(z, \rho), \partial E) \leq d(y, \partial E) = 10^{-3}\rho$. To prove that $d(B(z, \rho), \partial E) \geq 10^{-3}\rho$, observe that for any $\epsilon > 0$, $B(z, \rho) \subset B(z_j, \rho_j + \epsilon)$ for all j large enough. It follows that $d(B(z, \rho), \partial E) \geq 10^{-3}\rho_j - \epsilon$ for all j large enough. Letting first j tend to infinity and then ϵ tend to zero yields $d(B(z, \rho), \partial E) \geq 10^{-3}\rho$.

This procedure defines $\mathcal{F} = \{B_0, B_1, \ldots, B_k, \ldots\}$ inductively. By construction, properties (1) and (3) are satisfied. Let us show that property (2) is also satisfied. Fix $z \in E$. By continuity, there exists a $\rho > 0$ such that $d(B(z, \rho), \partial E) = 10^{-3}\rho$. Let k be the largest integer such that the ball $B_k \in \mathcal{F}$ has radius $r_k \geq \rho$. By construction, we must have $B(z, \rho) \cap (\bigcup_{i \leq k} B_i) \neq \emptyset$ because, if not, B_{k+1} must have radius $r_{k+1} \geq \rho$ and this contradicts the definition of k. Now, $B(z, \rho) \cap (\bigcup_{i \leq k} B_i) \neq \emptyset$ and $\rho \leq r_i$, $0 \leq i \leq k$, imply that $z \in \bigcup_{i \leq k} 2B_i$ as desired.

Finally, we prove property (4). Fix $\eta \in E$. Let $t = d(\eta, \partial E)$ so that $t \in (0, 2r)$. Let $B = B(z, \rho) \in \mathcal{F}$ such that $\eta \in 100B$. Then

$$10^3\rho = d(B, \partial E) \leq t + 101\rho$$
$$t = d(\eta, \partial E) \leq 101\rho + d(B, \partial E) \leq 10^4\rho.$$

Thus all the balls $B \in \mathcal{F}$ such that $10B$ contains η have radius at least $10^{-4}t$ and are all contained in $B(\eta, (101/899)t)$. As they are disjoint, (5.3.2) easily gives a bound on how many there are.

Let us now fix a covering $\mathcal{F}$ of E having the properties (1)–(4) above. There is a ball $B_x \in \mathcal{F}$ such that the center x of E belongs to $2B_x$. We call B_x the central ball of $\mathcal{F}$. Fix a ball $B \in \mathcal{F}$. Let η_B be the center of B and fix a distance-minimizing curve γ_B joining x to η_B.

Lemma 5.3.6 *For any $B \in \mathcal{F}$,*

$$d(\gamma_B, \partial E) \geq (1/2)d(B, \partial E) = (1/2)10^3 r(B).$$

In particular, any ball B' in $\mathcal{F}$ such that $2B'$ intersects γ_B has radius bounded below by

$$r(B') \geq (1/4)r(B).$$

Indeed, let $\zeta \in \gamma_B$ be such that $d(\zeta, \partial E) = d(\gamma_B, \partial E)$. Then

$$d(x, \zeta) + d(\zeta, \partial E) \geq R$$

and

$$d(\eta_B, \zeta) + d(\zeta, \partial E) \geq d(B, \partial E).$$

Moreover, $d(x, \zeta) + d(\zeta, \eta_B) = d(x, \eta_B)$. Hence

$$d(x, \eta_B) + 2d(\zeta, \partial E) \geq R + d(B, \partial E).$$

As $d(x, \eta_B) \leq R$, this yields

$$2d(\zeta, \partial E) = 2d(\gamma_B, \partial E) \geq d(B, \partial E)$$

which is the desired inequality.

Now, for any ball $B' \in \mathcal{F}$ such that $2B'$ intersects γ_B,

$$d(B', \partial E) \geq r(B') + d(\gamma_B, \partial E).$$

Thus $(10^3 - 1)r(B') \geq (1/2)10^3 r(B)$, which implies $r(B') \geq (1/4)r(B)$. This finishes the proof of Lemma 5.3.6.

Next we introduce an important notation. For any $B \in \mathcal{F}$, we now choose a string of balls in $\mathcal{F}$, call it

$$\mathcal{F}(B) = (B_0, B_1, \ldots, B_{\ell(B)-1}),$$

joining B_x to B with $B_0 = B_x$, $B_{\ell(B)-1} = B$ and with the property that $\overline{2B_i} \cap \overline{2B_{i+1}} \neq \emptyset$. Let us show that such a string exists. Let ξ_0 be the first point along γ_B (starting from x) which does not belong to $2B_0$ (recall that $x \in 2B_0$). Define B_1 to be one of the balls in $\mathcal{F}$ such that $2B_1$ contains ξ_0. Having constructed $B_0, B_1, \ldots, B_k$, let ξ_k be the first point along γ_B that does not belong to $\bigcup_0^k 2B_i$, and let B_{k+1} be one of the balls in $\mathcal{F}$ such that $2B_k$ contains ξ_k.

By (5.3.2), Lemma 5.3.6 and the fact that the balls in $\mathcal{F}(B)$ are disjoint, there are only finitely many balls B' in $\mathcal{F}$ that can intersect γ_B. In particular, $\mathcal{F}(B)$ is finite. It may well be that the last chosen ball in the above construction is not B. In this case, we simply add B as the last ball in $\mathcal{F}(B)$. Let us emphasize that the collection $\mathcal{F}(B)$ is finite but that we have no precise information on its cardinality $\ell(B)$.

Lemma 5.3.7 *For any $B \in \mathcal{F}$ and any two consecutive balls B_i, B_{i+1} in the string $\mathcal{F}(B)$,*

$$(1 + 10^{-2})^{-1} r(B_i) \leq r(B_{i+1}) \leq (1 + 10^{-2}) r(B_i)$$

and $B_{i+1} \subset 4B_i$. Moreover

$$\mu(4B_i \cap 4B_{i+1}) \geq c \max\{\mu(B_i), \mu(B_{i+1})\}.$$

As the balls $\overline{2B_i}$, $\overline{2B_{i+1}}$ intersect each other, one easily checks that B_i, B_{i+1} have comparable radii and that $B_{i+1} \subset 4B_i$ (this follows from the fact that each of these balls has radius equal to a small multiple of its distance from the boundary of E). Moreover, if $\xi \in \overline{2B_i} \cap \overline{2B_{i+1}}$ and $\rho = \min\{r(B_i), r(B_{i+1})\}$ then $B(\xi, \rho) \subset 4B_i \cap 4B_{i+1}$. Lemma 5.3.7 now follows from the doubling property (5.3.2) which shows that the balls B_i, B_{i+1}, $B(\xi, \rho)$ have comparable volume.

Lemma 5.3.8 *For any ball $B \in \mathcal{F}$ and any ball $A \in \mathcal{F}(B)$, $B \subset 10^4 A$.*

Let η_B be the center of B and η_A be the center of A. As $A \in \mathcal{F}(B)$, $2A \cap \gamma_B \neq \emptyset$. Thus, by Lemma 5.3.6,

$$r(A) \geq (1/4)r(B).$$

Now, let η'_A be a point in $2A \cap \gamma_B$. We have

$$d(\eta'_A, \eta_B) = d(x, \eta_B) - d(x, \eta'_A) \leq R - d(x, \eta'_A)$$

and

$$R \leq d(x, \eta'_A) + d(\eta'_A, \partial E).$$

Thus,

$$d(\eta'_A, \eta_B) \leq d(\eta'_A, \partial E) \leq (3 + 10^3)r(A)$$

and

$$d(\eta_A, \eta_B) \leq 2r(A) + d(\eta'_A, \eta_B)$$
$$\leq (5 + 10^3)r(A).$$

Thus

$$B \subset (9 + 10^3)A.$$

Lemma 5.3.9 *Under the hypotheses of Theorem 5.3.4 and assuming (as we may) that (5.3.1) holds with $\kappa = 2$, there exists a constant C such that for any $B \in \mathcal{F}$ and any consecutive balls B_i, B_{i+1} in $\mathcal{F}(B)$,*

$$|f_{4B_i} - f_{4B_{i+1}}| \leq C \frac{r(B_j)}{\mu(B_j)^{1/p}} \left(\int_{32B_j} |\nabla f|^p d\mu \right)^{1/p}.$$

For the proof, write

$$\mu(4B_i \cap 4B_{i+1})^{1/p} |f_{4B_i} - f_{4B_{i+1}}| = \left(\int_{4B_i \cap 4B_{i+1}} |f_{4B_i} - f_{4B_{i+1}}|^p d\mu \right)^{1/p}$$

$$\leq \left(\int_{4B_i \cap 4B_{i+1}} |f - f_{4B_i}|^p d\mu \right)^{1/p} + \left(\int_{4B_i \cap 4B_{i+1}} |f - f_{4B_{i+1}}|^p d\mu \right)^{1/p}$$

$$\leq \left(\int_{4B_i} |f - f_{4B_i}|^p d\mu \right)^{1/p} + \left(\int_{4B_{i+1}} |f - f_{4B_{i+1}}|^p d\mu \right)^{1/p}$$

$$\leq Cr(B_i) \left(\int_{8B_i} |\nabla f|^p d\mu \right)^{1/p} + Cr(B_{i+1}) \left(\int_{8B_{i+1}} |\nabla f|^p d\mu \right)^{1/p}. \quad (5.3.8)$$

The desired conclusion thus follows from (5.3.8) and Lemma 5.3.7 which shows that $r(B_i) \approx r(B_{i+1})$, $B_{i+1} \subset 4B_i$ and $\mu(4B_i \cap 4B_{i+1}) \approx \mu(B_i)$.

The next lemma extends Lemma 5.3.9 to the case of a non-trivial weight ϕ. We will use the following notation. For any $\phi \in \mathcal{M}(r, \alpha)$ and any $x \in M$, we let $E = B(x, r)$ and consider the Whitney covering $\mathcal{F}$ of E as above. Moreover, we set $\Phi(y) = \phi(d(x, y))$ as in Theorem 5.3.4.

Lemma 5.3.10 *Under the hypotheses of Theorem 5.3.4 and assuming (as we may) that (5.3.1) holds with $\kappa = 2$, there exists a constant C such that for any $B \in \mathcal{F}$ and any consecutive balls B_i, B_{i+1} in $\mathcal{F}(B)$,*

$$|f_{4B_i} - f_{4B_{i+1}}| \left(\frac{\Phi(B)}{\mu(B)}\right)^{1/p} \leq C \frac{r(B_j)}{\mu(B_j)^{1/p}} \left(\int_{32B_j} |\nabla f|^p \Phi d\mu\right)^{1/p}$$

where $\Phi(B) = \int_B \Phi d\mu$.

By Lemma 5.3.9 and the fact that Φ is essentially constant on $32B$ if $B \in \mathcal{F}$ we have

$$|f_{4B_i} - f_{4B_{i+1}}| \leq C \frac{r(B_j)}{\Phi(B_j)^{1/p}} \left(\int_{32B_j} |\nabla f|^p \Phi d\mu\right)^{1/p}.$$

By Lemma 5.3.6 and the properties of functions in $\mathcal{M}(r, \alpha)$, there exists a constant $c > 0$ such that $\Phi(B_i)/\mu(B_i) \geq c\Phi(B)/\mu(B)$ for all $B_i \in \mathcal{F}(B)$. The desired inequality follows.

5.3.4 A maximal inequality and an application

Let $f \in L^1 + L^\infty$. The maximal function $M_r f$ is defined by

$$M_r f(x) = \sup_{\substack{B:x\in B \\ r(B)<r}} \frac{1}{\mu(B)} \int_B |f| d\mu.$$

Theorem 5.3.11 *Fix $R > 0$ and assume that the doubling condition (5.3.2) is satisfied. Then for all $1 < p \leq \infty$ and $K \geq 1$, there exists $C = C(p, K)$ such that for all $0 < r < KR$,*

$$\forall f \in \mathcal{C}_0^\infty(M), \quad \|M_r f\|_p \leq C\|f\|_p.$$

If (5.3.2) holds with $R = \infty$, then one can take $r = \infty$.

This is obvious for $p = \infty$. By the Marcinkiewicz interpolation theorem, it suffices to show that M_r is of weak L^1 type. Thus it suffices to show that there exists C such that for any $f \in \mathcal{C}_0^\infty(M)$, $f \geq 0$, and $\lambda > 0$

$$\mu(E_\lambda) \leq C\lambda^{-1}\|f\|_1$$

where $E_\lambda = \{x : M_r f(x) > \lambda\}$.

Now, for any $x \in E_\lambda$ there exists a ball B_x of radius less than r such that

$$\int_{B_x} |f| d\mu \geq \lambda \mu(B_x).$$

Obviously, the balls B_x cover E_λ. By a well-known covering argument, one can extract from the family $\{B_x : x \in E_\lambda\}$ a sequence of balls (B_i) that are disjoint and such that $(5B_i)$ covers E_λ. See, e.g., [79, page 9]. By (5.3.1) and Lemma 5.2.7, it follows that

$$
\begin{aligned}
\mu(E_\lambda) &\leq \sum \mu(5B_i) \leq C \sum \mu(B_i) \\
&\leq C\lambda^{-1} \sum \int_{B_i} |f| d\mu \\
&\leq C\lambda^{-1} \int |f| d\mu.
\end{aligned}
$$

This finishes the proof of Theorem 5.3.11.

Lemma 5.3.12 *Fix $R > 0$ and assume that the doubling condition (5.3.2) is satisfied. Fix $K \geq 1$ and $1 \leq p < \infty$. Then there exists a constant $C = C(K, p)$ such that for any sequence $(B_i)_1^\infty$ of balls of radius at most R, and any sequence of non-negative numbers $(a_i)_1^\infty$,*

$$
\| \sum_i a_i \chi_{KB_i} \|_p \leq C \| \sum_i a_i \chi_{B_i} \|_p.
$$

Set $f = \sum_i a_i \chi_{KB_i}$, $g = \sum_i a_i \chi_{B_i}$. It suffices to show that for any $\phi \in C_0^\infty(M)$,

$$
\int f \phi d\mu \leq C \|g\|_p \|\phi\|_q
$$

where $1 = 1/p + 1/q$. Write

$$
\begin{aligned}
\int f \phi d\mu &= \sum_i a_i \int \chi_{KB_i}(x) \phi(x) dx \\
&= \sum_i a_i \mu(KB_i) \frac{1}{\mu(KB_i)} \int_{KB_i} \phi d\mu \\
&\leq C \sum_i a_i \mu(B_i) \frac{1}{\mu(KB_i)} \int_{KB_i} \phi d\mu.
\end{aligned}
$$

Now, for any $x \in B_i$,

$$
\frac{1}{\mu(KB_i)} \int_{KB_i} \phi d\mu \leq M_{KR} \phi(x).
$$

Hence

$$
\frac{1}{\mu(KB_i)} \int_{KB_i} \phi d\mu \leq \frac{1}{\mu(B_i)} \int_{B_i} M_{KR} \phi d\mu.
$$

It follows that

$$\int f\phi d\mu \;\leq\; C\sum_i a_i \int_{B_i} M_{KR}\phi d\mu$$

$$= \; C\int \sum_i a_i \chi_{B_i}(x) M_{KR}\phi(x) dx$$

$$\leq \; C\|g\|_p \|M_{KR}\phi\|_q.$$

As $1 \leq p < \infty$ implies $1 < q \leq \infty$, we can apply Theorem 5.3.11 which yields

$$\int f\phi d\mu \leq C'\|g\|_p \|\phi\|_q$$

as desired.

5.3.5 End of the proof of Theorem 5.3.4

We keep the notation introduced in Section 5.3.3. We also assume that the hypotheses of Theorem 5.3.4 are satisfied. Thus, ϕ is a function in $\mathcal{M}(r,\alpha)$ with $0 < r < R$ and $\Phi(y) = \phi(d(x,y))$ for some fixed $x \in M$. Moreover, $E = B(x,r)$ and $\mathcal{F}$ is a Whitney-type covering of E as in Section 5.3.3. Recall that $\mathcal{F}$ contains a so-called central ball B_x with the property that $x \in 2B_x$. Any ball $B \in \mathcal{F}$ comes equipped with a finite string of balls of $\mathcal{F}$ which is denoted by $\mathcal{F}(B) = (B_0, \dots, B_{\ell(B)-1})$. This string has a number of specific properties explained in Section 5.3.3. In particular, $B_0 = B_x$, $B_{\ell(B)-1} = B$.

Recall that $\Phi(B) = \int_B \Phi d\mu$. As $E = \bigcup_{B\in\mathcal{F}} 2B$ and Φ has support in E, we have

$$\int |f - f_{4B_x}|^p \Phi d\mu \tag{5.3.9}$$

$$\leq \; \sum_{B\in\mathcal{F}} \int_{2B} |f - f_{4B_x}|^p \Phi d\mu$$

$$\leq \; 2^p \sum_{B\in\mathcal{F}} \int_{4B} (|f - f_{4B}|^p + |f_{4B} - f_{4B_x}|^p)\Phi d\mu$$

$$\leq \; 2^p \sum_{B\in\mathcal{F}} \int_{4B} |f - f_{4B}|^p \Phi d\mu + \sum_{B\in\mathcal{F}} |f_{4B} - f_{4B_x}|^p \Phi(4B). \tag{5.3.10}$$

By hypothesis, Φ is essentially constant on the balls $4B$, $B \in \mathcal{F}$. Thus (5.3.1) implies

$$\int_{4B} |f - f_{4B}|^p \Phi d\mu \leq P_0 r(4B)^p \int_{8B} |\nabla f|^p \Phi d\mu.$$

Hence

$$\sum_{B \in \mathcal{F}} \int_{4B} |f - f_{4B}|^p \Phi d\mu \leq P_0 \sum_{B \in \mathcal{F}} r(4B)^p \int_{8B} |\nabla f|^p \Phi d\mu$$

$$\leq 4^p P_0 \, r^p \int_E |\nabla f|^p \Phi d\mu \qquad (5.3.11)$$

where the last inequality uses (5.3.7) and the fact that $8B \subset E$ for any $B \in \mathcal{F}$.

We now have to bound

$$I = \sum_{B \in \mathcal{F}} |f_{4B} - f_{4B_x}|^p \Phi(4B) \leq C \sum_{B \in \mathcal{F}} \int |f_{4B} - f_{4B_x}|^p \frac{\Phi(B)}{\mu(B)} \chi_B d\mu.$$

To this end, recall that $B_x = B_0$, $B = B_{\ell(B)-1}$ and, using the string of balls $\mathcal{F}(B) = (B_0, B_1, \ldots, B_{\ell-1})$ and Lemma 5.3.10, write

$$|f_{4B} - f_{4B_0}| \left(\frac{\Phi(B)}{\mu(B)} \right)^{1/p} \leq \sum_0^{\ell(B)-1} |f_{4B_i} - f_{4B_{i+1}}| \left(\frac{\Phi(B)}{\mu(B)} \right)^{1/p}$$

$$\leq C \sum_0^{\ell(B)-1} \frac{r(B_i)}{\mu(B_i)^{1/p}} \left(\int_{32B_i} |\nabla f|^p \Phi d\mu \right)^{1/p}.$$

By Lemma 5.3.8, the ball B is contained in $10^4 B_i$ for any $B_i \in \mathcal{F}(B)$. Thus, the last inequality yields

$$|f_{4B} - f_{4B_0}| \left(\frac{\Phi(B)}{\mu(B)} \right)^{1/p} \chi_B \leq C \sum_{A \in \mathcal{F}} \frac{r(A)}{\mu(A)^{1/p}} \left(\int_{32A} |\nabla f|^p \Phi d\mu \right)^{1/p} \chi_{10^4 A} \chi_B.$$

As $\sum_{B \in \mathcal{F}} \chi_B \leq 1$ (the balls in $\mathcal{F}$ are disjoint), we get

$$\sum_{B \in \mathcal{F}} |f_{4B} - f_{4B_0}|^p \frac{\Phi(B)}{\mu(B)} \chi_B \leq C \left| \sum_{A \in \mathcal{F}} \frac{r(A)}{\mu(A)^{1/p}} \left(\int_{32A} |\nabla f|^p \Phi d\mu \right)^{1/p} \chi_{10^5 A} \right|^p.$$

By Lemma 5.3.12 and since the balls in $\mathcal{F}$ are disjoint, this yields

$$\int \sum_{B \in \mathcal{F}} |f_{4B} - f_{4B_0}|^p \frac{\Phi(B)}{\mu(B)} \chi_B d\mu$$

$$\leq C' \int \left| \sum_{A \in \mathcal{F}} \frac{r(A)}{\mu(A)^{1/p}} \left(\int_{32A} |\nabla f|^p \Phi d\mu \right)^{1/p} \chi_A \right|^p d\mu$$

$$= C' \int \sum_{A \in \mathcal{F}} \frac{r(A)^p}{\mu(A)} \left(\int_{32A} |\nabla f|^p \Phi d\mu \right) \chi_A d\mu$$

$$\leq C' r^p \sum_{A \in \mathcal{F}} \int_{32A} |\nabla f|^p \Phi d\mu \leq C'' r^p \int |\nabla f|^p \Phi d\mu. \qquad (5.3.12)$$

For the last step, we have used (5.3.7) and the fact that $32A \subset E$ for all $A \in \mathcal{F}$.

To conclude the proof of Theorem 5.3.4, it now suffices to use (5.3.10), (5.3.11) and (5.3.12). Indeed, these inequalities yield

$$\int |f - f_{4B_x}|^p \Phi d\mu \leq Cr^p \int |\nabla f|^p \Phi d\mu.$$

As this implies $|f_\Phi - f_{4B_x}|^p \int \Phi d\mu \leq Cr^p \int |\nabla f|^p \Phi d\mu$, we conclude that

$$\int |f - f_\Phi|^p \Phi d\mu \leq Cr^p \int |\nabla f|^p \Phi d\mu$$

as desired.

5.4   Harnack inequalities and applications

5.4.1   An inequality for $\log u$

Recall that $d\bar{\mu}$ denotes the natural product measure on $\mathbb{R} \times M$: $d\bar{\mu} = dt \times d\mu$.

Lemma 5.4.1 *Fix $0 < R \leq \infty$. Fix $\tau > 0$ and $\delta, \eta \in (0,1)$. Assume that (5.3.1), (5.3.2) are satisfied. For any real s, any r with $0 < r < R$, any ball B of radius r, and any positive function u such that $(\partial_t + \Delta)u \geq 0$ in $Q = (s - \tau r^2, s) \times B$, there is a constant $c = c(u, \eta)$ such that, for all $\lambda > 0$,*

$$\bar{\mu}\left(\{(t,z) \in K_+ : \log u < -\lambda - c\}\right) \leq Cr^2\mu(B)\lambda^{-1}$$

and

$$\bar{\mu}\left(\{(t,z) \in K_- : \log u > \lambda - c\}\right) \leq Cr^2\mu(B)\lambda^{-1}$$

where $K_+ = (s - \eta\tau r^2, s) \times \delta B$ and $K_- = (s - r^2, s - \eta\tau r^2) \times \delta B$. Here the constant C is independent of $\lambda > 0$, u, s and the ball B of radius $r \in (0, R)$.

For the proof, we assume that $\tau = 1$. Note that δ and η play somewhat different roles here. The parameter δ is used to stay away from the boundary of the ball B. The parameter η is used to define a fixed point $s' = s - \eta r^2$ in the interval $(s - r^2, s)$, away from $s - r^2$ and s.

Let us first observe that (by changing δ) we can assume that u is a supersolution in $(s - r^2, s) \times B'$ where B' is a concentric ball larger than $B = B(x, r)$. We set $w = -\log u$. Then, for any non-negative function $\psi \in \mathcal{C}_0^\infty(B')$, we have

$$\partial_t \int \psi^2 w \, d\mu \leq \int \psi^2 u^{-1} \Delta u \, d\mu = \int \left[-\psi^2 |\nabla w|^2 + 2\psi \nabla w \cdot \nabla \psi\right] d\mu.$$

Using $2|ab| \leq (\frac{1}{2}a^2 + 2b^2)$, we get

$$\partial_t \int \psi^2 w + \frac{1}{2} \int |\nabla w|^2 \psi^2 \leq 2\|\nabla\psi\|_\infty^2 \mu(\mathrm{supp}(\psi)). \qquad (5.4.1)$$

Here, we choose $\psi(z) = (1 - \rho(x,z)/r)_+$ where x is the center of B and r its radius (ψ is not smooth, but it can easily be approximated by non-negative functions in $\mathcal{C}_0^\infty(B')$). In the notation of Theorem 5.3.4, we have $\psi^2(y) = \Phi(y)$ with $\phi(t) = (1 - t/r)_+^2$, $t > 0$. Thus, the weighted Poincaré inequality of Theorem 5.3.4 with weight $\Phi_x = \psi^2$ yields

$$\int |w - W|^2 \psi^2 d\mu \leq A_0 r^2 \int |\nabla w|^2 \psi^2 d\mu$$

with

$$W = \int w\psi^2 d\mu \Big/ \int \psi^2 d\mu.$$

Setting

$$V = \mu(B),$$

this and (5.4.1) give

$$\partial_t W + (A_1 r^2 V)^{-1} \int_{\delta B} |w - W|^2 d\mu \leq A_2 r^{-2}$$

for some constant $A_1, A_2 > 0$. Rewrite this inequality as

$$\partial_t \overline{W} + (A_1 r^2 V)^{-1} \int_{\delta B} |\overline{w} - \overline{W}|^2 d\mu \leq 0 \qquad (5.4.2)$$

where

$$\overline{w}(t,z) = w(t,z) - A_2 r^{-2}(t - s')$$
$$\overline{W}(t) = W(t) - A_2 r^{-2}(t - s')$$

with $s' = s - \eta r^2$.

Now, set $c(u) = \overline{W}(s')$, and

$$\Omega_t^+(\lambda) = \{z \in \delta B : \overline{w}(t,z) > c + \lambda\}$$

$$\Omega_t^-(\lambda) = \{z \in \delta B : \overline{w}(t,z) < c - \lambda\}.$$

Then, if $t > s'$,

$$\overline{w}(t,z) - \overline{W}(t) \geq \lambda + c - \overline{W}(t) > \lambda$$

in $\Omega_t^+(\lambda)$, because $c = \overline{W}(s')$ and $\partial_t \overline{W} \leq 0$. Using this in (5.4.2), we obtain

$$\partial_t \overline{W}(t) + (A_1 r^2 V)^{-1} |\lambda + c - \overline{W}(t)|^2 \mu(\Omega_t^+(\lambda)) \leq 0$$

or, equivalently,

$$-A_1 r^2 V \, \partial_t \left(|\lambda + c - \overline{W}(t)|^{-1}\right) \geq \mu(\Omega_t^+(\lambda)).$$

Integrating from s' to s, we obtain

$$\bar{\mu}\left(\{(t,z) \in K_+ : \overline{w}(t,z) > c + \lambda\}\right) \leq A_1 r^2 V \lambda^{-1}$$

and, returning to $-\log u = w = \overline{w} + A_2 r^{-2}(t - s')$,

$$\bar{\mu}\left(\{(t,z) \in K_+ : \log u(t,z) + A_2 r^{-2}(t - s') < -\lambda - c\}\right) \leq A_1 r^2 V \lambda^{-1}.$$

Finally,

$$\begin{aligned}
&\bar{\mu}\left(\{(t,z) \in K_+ : \log u(t,z) < -\lambda - c\}\right) \\
&\leq \quad \bar{\mu}\left(\{(t,z) \in K_+ : \log u(t,z) + A_2 r^{-2}(t - s') < -(\lambda/2) - c\}\right) \\
&\quad + \bar{\mu}\left(\{(t,z) \in K_+ : A_2 r^{-2}(t - s') > \lambda/2\}\right) \\
&\leq \quad A_3 r^2 V \lambda^{-1}.
\end{aligned}$$

This proves the first inequality in Lemma 5.4.1. Working with Ω_t^- instead of Ω_t^+, we obtain the second inequality by a similar argument.

5.4.2 Harnack inequality for positive supersolutions

The following theorem states that positive supersolutions satisfy a weak form of Harnack inequality. For any fixed $\tau > 0$, $\delta \in (0,1)$ and $x \in M$, $s, r > 0$ define

$$\begin{aligned}
Q_- &= (s - (3+\delta)\tau r^2/4, s - (3-\delta)\tau r^2/4) \times \delta B \\
Q'_- &= (s - \tau r^2, s - (3-\delta)\tau r^2/4) \times \delta B \\
Q_+ &= (s - (1+\delta)\tau r^2/4, s) \times \delta B.
\end{aligned}$$

Recall also that $Q = Q(\tau, x, s, r) = (s - \tau r^2) \times B(x, r)$.

Theorem 5.4.2 *Fix $\tau > 0$, $0 < \delta < 1$ and $0 < R \leq \infty$. Assume that (5.3.1), (5.3.2) are satisfied for this R with $p = 2$. Let C and $\nu > 2$ be such that (5.3.6) is satisfied with $p = 2$ (these exist by Theorem 5.3.3). Fix $p_0 \in (0, 1 + 2/\nu)$. Then there exists a constant A such that, for $x \in M$, $s \in \mathbb{R}$, $0 < r < R$ and any positive function u satisfying $(\partial_t + \Delta)u \geq 0$ in $Q = (s - \tau r^2, s) \times B(x, r)$, we have*

$$\left(\frac{1}{\mu(Q'_-)} \int_{Q'_-} u^{p_0} d\bar{\mu}\right)^{1/p_0} \leq A \inf_{Q_+}\{u\}.$$

For simplicity, we assume for the proof that $\tau = 1$.

Fix a non-negative supersolution u. Let $c(u)$ be the constant given by Lemma 5.4.1 applied to u with $\eta = 1/2$. Set $v = e^c u$. Set also

$$U = (s - r^2, s - (1/2)r^2) \times B, \quad U_\sigma = (s - r^2, s - (3-\sigma)r^2/4) \times \sigma B.$$

By Theorem 5.2.17, we have

$$\left(\int_{U_{\sigma'}} v^{p_0} d\bar{\mu}\right)^{1/p_0} \leq \left[\frac{A(p_0, \nu)C^{1+\nu}}{(\sigma - \sigma')^{4+2\nu}\bar{\mu}(U)}\right]^{1/p - 1/p_0} \left(\int_{U_\sigma} v^p d\bar{\mu}\right)^{1/p}$$

for all $0 < \delta < \sigma' < \sigma < 1$ and $0 < p < p_0(1 + 2/\nu)^{-1}$. By Lemma 5.4.1, we have

$$\bar{\mu}(\log v > \lambda) \leq C\bar{\mu}(U).$$

Thus we can apply Lemma 2.2.6 to conclude that $\int_{U_\delta} v^{p_0} d\bar{\mu} \leq A_1 \bar{\mu}(U)$, that is

$$\left(\frac{1}{\bar{\mu}(Q'_-)} \int_{Q'_-} (e^c u)^{\alpha_0} d\bar{\mu}\right)^{1/p_0} \leq A'_1. \tag{5.4.3}$$

Set now $v = e^{-c} u^{-1}$, where $c = c(u)$ is the same constant as above, given by Lemma 5.4.1 applied to u with $\eta = 1/2$. This time, set

$$U = (s - (1/2)r^2, s) \times B, \quad U_\sigma = (s - (1 + \sigma)r^2/4, s) \times \sigma B.$$

By Theorem 5.2.16, we have

$$\sup_{U_{\sigma'}}\{v^p\} \leq \frac{A(\nu)C^{\nu/2}}{(\sigma - \sigma')^{2+\nu}\bar{\mu}(U)} \int_{U_\sigma} v^p d\bar{\mu}$$

for all $0 < \delta < \sigma' < \sigma < 1$ and $0 < p < \infty$. By Lemma 5.4.1, we also have

$$\bar{\mu}(\log v > \lambda) \leq C\bar{\mu}(U).$$

Thus we can again apply Lemma 2.2.6 to conclude that $\sup_{U_\delta}\{v\} \leq A_2\bar{\mu}(U)$, that is

$$\sup_{Q_+}\{(e^c u)^{-1}\} \leq A_2. \tag{5.4.4}$$

Multiplying (5.4.3), (5.4.4) together, we obtain

$$\left(\frac{1}{\bar{\mu}(Q'_-)} \int_{Q'_-} u^{p_0} d\bar{\mu}\right)^{1/p_0} \leq A \inf_{Q_+}\{u\}$$

which is the desired inequality.

5.4.3   Harnack inequalities for positive solutions

In this section, we describe different forms of Harnack inequalities for positive solutions of the heat diffusion equation

$$(\partial_t + \Delta)u = 0.$$

The basic assumption is that there exists $0 < R \leq \infty$ such that (5.3.1), (5.3.2) are satisfied with $p = 2$, that is, M satisfies a scale-invariant Poincaré inequality (5.3.1) with $p = 2$ and the doubling condition (5.3.2), up to radius R.

Theorem 5.4.3 *Fix* $\tau > 0$, $0 < \delta < 1$ *and* $0 < R \leq \infty$. *Assume that* (5.3.1), (5.3.2) *are satisfied for this* R *with* $p = 2$. *Then there exists a constant* A *such that, for* $x \in M$, $s \in \mathbb{R}$, $0 < r < R$ *and any positive function* u *satisfying* $(\partial_t + \Delta)u = 0$ *in* $Q = (s - \tau r^2, s) \times B(x, r)$, *we have*

$$\sup_{Q_-}\{u\} \leq A \inf_{Q_+}\{u\}$$

where

$$\begin{aligned} Q_- &= (s - (3+\delta)\tau r^2/4, s - (3-\delta)\tau r^2/4) \times \delta B \\ Q_+ &= (s - (1+\delta)\tau r^2/4, s) \times \delta B. \end{aligned}$$

This follows immediately from Theorems 5.4.2 and 5.2.9.

Next we give a useful corollary of this inequality.

Corollary 5.4.4 *Fix* $0 < R \leq \infty$. *Assume that* (5.3.1), (5.3.2) *are satisfied. There exists a constant* A *such that the following inequality holds. Let* γ *be a continuous curve of length* d *joining two points* $x, y \in M$. *Let* Γ_ρ *be a* ρ-*neighborhood of* γ, $\rho > 0$. *Let* u *be a non-negative solution of* $(\partial_t + \Delta)u = 0$ *in* $(0, T) \times \Gamma_\rho$, $T > 0$ *and let* $0 < s < t < T$. *Then*

$$\log \frac{u(s, x)}{u(t, y)} \leq A \left(1 + \frac{t-s}{R^2} + \frac{t-s}{s} + \frac{t-s}{\rho^2} + \frac{d^2}{t-s} \right).$$

Connect the points x, y by a string of k balls $B_0, \ldots, B_{k-1}$ of radius r and centers $x_0, \ldots, x_{k-1}$, with $x_0, \ldots, x_{k-1} \in \gamma$, $x_0 = x$, and $x_{i+1} \in \overline{B}_i$, $0 \leq i \leq k - 1$ with $x_k = y$. This is possible as soon as

$$kr \geq d. \tag{5.4.5}$$

The values of r and k are to be chosen later. Let $t_0 = s$, $t_i = s + r^2 i$, $0 \leq i \leq k$. Now choose r to satisfy the following conditions:

(i) $r^2 = (t - s)/k$ so that $t_k = t$. Note that this implies $t - s \geq r^2$.

(ii) $r \leq R$, $r^2 \leq s$ and $r \leq 2\rho$ so that u is a solution of $(\partial_t + \Delta)u = 0$ in each of the cylinders $(t_i - r^2, t_{i+1}) \times 2B_i$, $0 \leq i \leq k - 1$.

Then, applying Theorem 5.4.3 successively in $(t_i - r^2, t_i) \times 2B_i$, $0 \leq i \leq k-1$, we obtain

$$\begin{aligned} u(t_0, x_0) &\leq A_0 u(t_1, x_1) \leq A_0^2 u(t_2, x_2) \leq \cdots \\ &\leq A_0^k u(t_k, x_k), \end{aligned}$$

that is,

$$u(s, x) \leq A_0^k u(t, y).$$

Now, (5.4.5) is satisfied if $k \geq d^2/(t-s)$ because $kr = \sqrt{k(t-s)}$ by (i). Similarly, (ii) is satisfied as soon as $k \geq (t-s) \max\{1/R^2, 1/s, 1/2\rho\}$. Thus we can choose k of order

$$1 + \frac{t-s}{R^2} + \frac{t-s}{s} + \frac{t-s}{\rho^2} + \frac{d^2}{t-s}.$$

This gives the desired inequality.

The next result improves the dependence on t/s in the inequality of Corollary 5.4.4 when u is a solution on $(0, T) \times M$ and (5.3.1), (5.3.2) are satisfied with $R = \infty$ and $p = 2$.

Corollary 5.4.5 *Assume that (5.3.1), (5.3.2) are satisfied with $R = \infty$ and $p = 2$. There exists a constant A such that the following inequality holds. Let $x, y \in M$. Let u be a non-negative solution of $(\partial_t + \Delta)u = 0$ in a $(0, T) \times M$, $T > 0$ and let $0 < s < t < T$. Then*

$$\log \frac{u(s, x)}{u(t, y)} \leq A \left(1 + \frac{d(x, y)^2}{t-s} + \log \frac{t}{s} \right).$$

If $s < t < 4s$, Corollary 5.4.4 yields

$$\log \frac{u(s, x)}{u(t, y)} \leq A_0 \left(1 + \frac{d(x, y)^2}{t-s} \right). \qquad (5.4.6)$$

If $t > 4s$, consider $k \geq 1$ such that

$$2^{k+1}s \leq t < 2^{k+2}s$$

and set

$$t_i = 2^i s, \quad 0 < i \leq k+1.$$

Then, by (5.4.6),

$$\log \frac{u(t_k, x)}{u(t, y)} \leq 2A_0 \left(1 + \frac{d(x, y)^2}{t-s} \right)$$

because $t_k < t < 2t_k$ and $t - t_k \geq t/2 \geq (t-s)/2$. Moreover, by (5.4.6),

$$\log \frac{u(t_{i-1}, x)}{u(t_i, x)} \leq A_0, \quad 1 \leq i \leq k,$$

because $t_i < t_{i+1} = 2t_i$. Thus

$$\log \frac{u(t_0, x)}{u(t_k, x)} \leq A_0 k.$$

As $t_0 = s$ and k is of order $\log(t/s)$, it follows that

$$\log \frac{u(s,x)}{u(t,y)} \leq A\left(1 + \frac{d(x,y)^2}{t-s} + \log \frac{t}{s}\right)$$

as desired.

The next corollary deals with global solutions on $(0,\infty) \times M$ and follows directly from the previous results.

Corollary 5.4.6 *Fix* $0 < R \leq \infty$. *Assume that* (5.3.1), (5.3.2) *are satisfied for this* R *with* $p = 2$.

(1) *If* $R = \infty$, *there exist constants* A *and* $a > 0$ *such that, for any* $x, y \in M$, *any* $0 < s < t < \infty$ *and any non-negative solution of* $(\partial_t + \Delta)u = 0$ *in* $(0,\infty) \times M$, *we have*

$$u(s,x) \leq u(t,y) \left(\frac{t}{s}\right)^a \exp\left(A\left(1 + \frac{d(x,y)^2}{t-s}\right)\right).$$

(2) *If* $R < \infty$, *we still have*

$$u(s,x) \leq u(t,y) \exp\left(A\left(1 + \frac{t-s}{R^2} + \frac{d(x,y)^2}{t-s}\right)\right).$$

5.4.4 Hölder continuity

One of the important consequences of the Harnack inequality of Theorem 5.4.3 is that it provides a quantitative Hölder continuity estimate for solutions of $(\partial_t + \Delta)u = 0$.

Theorem 5.4.7 *Fix* $0 < R \leq \infty$. *Fix* $\tau > 0$ *and* $\delta \in (0,1)$. *Assume that* (5.3.1), (5.3.2) *are satisfied for this* R *with* $p = 2$. *Then there exist* $\alpha \in (0,1)$ *and* $A > 0$ *such that any solution* u *of* $(\partial_t + \Delta)u = 0$ *in* $Q = (s - \tau r^2, s) \times B(x,r)$, $x \in M$, $0 < r < R$, *satisfies*

$$\sup_{(y,t),(y',t')\in Q_\delta} \left\{\frac{|u(y,t) - u(y',t')|}{[|t-t'|^{1/2} + d(y,y')]^\alpha}\right\} \leq \frac{A}{r^\alpha} \sup_Q \{|u|\}.$$

As usual, we assume that $\tau = 1$. Let us start with a simple consequence of Theorem 5.4.3. Fix $\sigma > 0$, $\rho \in (0,R)$, $z \in M$. Set

$$\begin{aligned}
W &= (\sigma - \rho^2, \sigma) \times B(z, \rho), \\
W_- &= (\sigma - \rho^2, \sigma - (3/4)\rho^2) \times B(z, (1/2)\rho), \\
W_+ &= (\sigma - (1/4)\rho^2, \sigma) \times B(z, (1/2)\rho).
\end{aligned}$$

Then for any non-negative solution v of $(\partial_t + \Delta)v = 0$ in W

$$\frac{1}{\bar{\mu}(W_-)} \int_{W_-} v \, d\bar{\mu} \leq \max_{W_-}\{v\} \leq A \min_{W_+}\{v\}. \tag{5.4.7}$$

Now, given a solution u, not necessarily non-negative, let M_u, m_u be the (essential) maximum and minimum of u in W. Similarly, let M_u^+, m_u^+ be the maximum and minimum of u in W_+. Define also

$$\mu_u^- = \frac{1}{\bar{\mu}(W_-)} \int_{W_-} v d\bar{\mu}.$$

Applying (5.4.7) to the non-negative solutions $M_u - u$, $u - m_u$, we obtain

$$M_u - \mu_u^- \leq A(M_u - M_u^+)$$
$$\mu_u^- - m_u \leq A(m_u^+ - m_u).$$

It follows that

$$(M_u - m_u) \leq A(M_u - m_u) - A(M_u^+ - m_u^+).$$

Thus the oscillations

$$\omega(u, W) = M_u - m_u$$
$$\omega(u, W_+) = M_u^+ - m_u^+$$

of u over W and W_+ satisfy

$$\omega(u, W^+) \leq \theta \, \omega(u, W) \tag{5.4.8}$$

with $\theta = 1 - 1/A \in (0, 1)$ (note that $A > 1$).

Now, referring to Theorem 5.4.7, consider $(y, t), (y', t') \in Q_\delta$. We can assume without loss of generality that $t \geq t'$. Let

$$\rho = 2 \max\{d(y, y'), \sqrt{t - t'}\}.$$

Then (y', t') belongs to $W_0 = (t - \rho^2, t) \times B(y, \rho)$. For $i \geq 1$, define $\rho_i = 2\rho_{i-1}$, $\rho_0 = \rho$, and set $W_i = (t - \rho_i^2, t) \times B(y, \rho_i)$. Then, according to the notation adopted for (5.4.7), we have

$$(W_i)_+ = W_{i-1}.$$

Thus, as long as W_i is contained in Q, (5.4.8) yields

$$\omega(u, W_{i-1}) \leq \theta \, \omega(u, W_i)$$

and

$$\omega(u, W_0) \leq \theta^i \, \omega(u, Q).$$

Consider two cases. If $\rho \leq (1 - \delta) r$, let k the integer such that

$$2^k \leq (1 - \delta) r/\rho < 2^{k+1}.$$

Then, as $(y, t) \in Q_\delta$, it follows that

$$
\begin{aligned}
W_k &= (t - 4^k \rho^2, t) \times B(y, 2^k \rho) \\
&\subset (t - (1 - \delta)^2 r^2, t) \times B(y, (1 - \delta) r) \\
&\subset (s - r^2, s) \times B(x, r) = Q.
\end{aligned}
$$

Thus

$$
\omega(u, W_0) \le \theta^k \, \omega(u, Q) \le \theta^{-1} (1 - \delta)^{-\alpha} (\rho/r)^\alpha \, \omega(u, Q)
$$

with $\alpha = \log_2 \theta$. In particular,

$$
\frac{|u(y, t) - u(y', t')|}{[|t - t'|^{1/2} + d(y, y')]^\alpha} \le \frac{A_\delta}{r^\alpha} \sup_Q \{|u|\}
$$

as desired.

The second case is trivial: if $\rho \ge (1 - \delta) r$, then the last inequality obviously holds. This ends the proof of Theorem 5.4.7.

Applying Theorem 5.4.7 to the heat kernel yields the following result.

Theorem 5.4.8 *Assume that* (5.3.1), (5.3.2) *are satisfied with* $p = 2$ *and* $R = \infty$. *Then there exist* $\alpha \in (0, 1)$ *and* $A > 0$ *such that*

$$
|h(t, x, y) - h(t', x, z)| \le A \left(\frac{|t - t'|^{1/2} + d(y, z)}{t^{1/2}} \right)^\alpha h(2t, x, y)
$$

for all $x \in M$, $t, t' > 0$ *and* $z, y \in M$, $d(z, y)^2 \le t$, $0 < |t - t'| < t/2$.

5.4.5 Liouville theorems

Let M be a complete Riemannian manifold. One says that M has the strong Liouville property if any solution u of the Laplace equation $\Delta u = 0$ on M which is bounded below (or above) is a constant. Usually, the weak Liouville property refers to the same result for *bounded* solutions. In the classical case of harmonic functions in $\mathbb{R}^n$, the strong (and weak) Liouville property is satisfied and one can also prove that a harmonic function satisfying $\lim_{r \to \infty} \frac{1}{r} \sup_{B(0,r)} \{|u|\} = 0$, i.e., having sublinear growth, must be constant. These Liouville properties are somewhat subtle properties. In particular, T. Lyons [58] proved that the strong and weak Liouville properties are not stable under quasi-isometric changes of metrics. It is well known that Liouville-type properties follow from Harnack-type inequalities. Indeed, assume that any non-negative solution u of $\Delta u = 0$ in a ball B satisfies the elliptic Harnack inequality

$$
\sup_{(1/2)B} \{u\} \le C \inf_{(1/2)B} \{u\}
$$

where C is independent of u and B. Let u be a solution of $\Delta u = 0$ in M which is bounded below. Let $m(u) = \inf_M \{u\}$. Applying the Harnack inequality above in the ball $2B$ and to the function $v = u - m(u)$, we find that

$$\sup_B \{u - m(u)\} \leq C \inf_B \{u - m(u)\}.$$

As the radius of B tends to infinity, $\inf_B \{u - m(u)\}$ tends to zero and we conclude that $u = m(u)$ is constant. Of course, the parabolic Harnack inequality of Theorem 5.4.3 implies the elliptic version used in the argument above. Thus we have proved the first assertion of the following theorem.

Theorem 5.4.9 *Assume that M is a complete Riemannian manifold satisfying* (5.3.1), (5.3.2) *with $p = 2$ and $R = \infty$. Then M has the strong Liouville property. Moreover, there exists an $\alpha > 0$ such that any function u such that $\Delta u = 0$ in M and*

$$\lim_{r \to \infty} \frac{1}{r^\alpha} \sup_{B(o,r)} \{|u|\} = 0$$

must be constant. Here o is some fixed reference point in M.

Only the second assertion still needs to be proved. Let α be as given by Theorem 5.4.7. Let u be a function such that $\Delta u = 0$ in M and

$$\lim_{r \to \infty} \frac{1}{r^\alpha} \sup_{B(o,r)} \{|u|\} = 0.$$

Fix some $x \in M$ and y such that $d(x, y) \leq 1$. Applying Theorem 5.4.7 to u in a ball $B_R = B(0, R)$ with R so large that $x, y \in (1/2)B_R$, we find that

$$|u(x) - u(y)| \leq C R^{-\alpha} \sup_{B_R} \{|u|\}.$$

As this holds for all R large enough, we can let R tend to infinity to obtain that $|u(x) - u(y)| = 0$. As x, y with $d(x, y) \leq 1$ are arbitrary and M is connected, u must be constant.

Let us conclude this section by pointing to a recent advance in this direction due to T. Colding and W. Minicozzi II [17]. They proved that, for complete Riemannian manifolds as in Theorem 5.4.9 above and for *any* $\alpha > 0$, the linear space of solutions of $\Delta u = 0$ with the growth property that

$$\sup_{r \geq 1} \frac{1}{r^\alpha} \sup_{B(o,r)} \{|u|\} < \infty$$

is finite dimensional. See also [53, 54].

5.4.6   Heat kernel lower bounds

The Harnack inequalities of Section 5.4.3 easily yield heat kernel lower bounds. First, we have the following on-diagonal lower bound.

Theorem 5.4.10 *Fix $0 < R \leq \infty$. Assume that (5.3.1), (5.3.2) are satisfied for this R with $p = 2$. Then the heat kernel on M satisfies*

$$h(t, x, x) \geq \frac{c}{V(x, \sqrt{t})}$$

for all $x \in M$ and all $0 < t < R^2$.

Fix $0 < t < R^2$. Let $B = B(x, \sqrt{t})$. Let ϕ be a smooth function such that $0 \leq \phi \leq 1$, $\phi = 1$ on B and $\phi = 0$ on $M \setminus 2B$. Define

$$u(t, y) = \begin{cases} H_t\phi(y) & \text{if } t > 0 \\ \phi(y) & \text{if } t \leq 0. \end{cases}$$

Obviously, this function satisfies

$$(\partial_t + \Delta)u = 0$$

on $(-\infty, +\infty) \times B$. Indeed, by local estimates, one has

$$\lim_{t \to 0} \frac{H_t\phi - \phi}{t} = -\Delta\phi = 0$$

pointwise in B. Applying Theorem 5.4.3, first to u, and then to the heat kernel $(s, y) \mapsto h(s, x, y)$, we get

$$
\begin{aligned}
1 = u(0, x) &\leq Au(t/2, x) \\
&= A \int h(t/2, x, y)\phi(y)dy \\
&\leq A \int_{2B} h(t/2, x, y)dy \\
&\leq A^2\mu(2B)h(t, x, x).
\end{aligned}
$$

This gives

$$h(t, x, x) \geq A^{-2}V(x, \sqrt{t})^{-1}$$

as desired.

Theorem 5.4.11 *Fix $0 < R \leq \infty$. Assume that (5.3.1), (5.3.2) are satisfied for this R with $p = 2$. There exists a constant A such that, for any $x, y \in M$ and any $0 < t < \infty$ the heat kernel $h(t, x, y)$ satisfies*

$$h(t, x, y) \geq h(t, x, x)\exp\left(-A\left(1 + \frac{t}{R^2} + \frac{d(x, y)^2}{t}\right)\right).$$

Moreover, there exists $a > 0$ such that, for all $x, y \in M$ and $0 < t < R^2$,

$$h(t, x, y) \geq \frac{a}{V(x, \sqrt{t})}\exp\left(-A\frac{d(x, y)^2}{t}\right).$$

Apply Corollary 5.4.6(2) to $u(t, y) = h(t, x, y)$ with x fixed and $s = t/2$. This gives the first stated inequality because $h(t, x, x)$ is non-increasing. The second inequality then follows from Theorem 5.4.10.

5.4.7   Two-sided heat kernel bounds

It might be useful to gather in one place different heat kernel estimates that have been obtained so far in the important case where there exists a constant D_0 such that the volume growth function $V(x, r)$ has the doubling property

$$\forall\, x \in M, \ \forall\, r > 0, \quad V(x, 2r) \le D_0 V(x, r) \tag{5.4.9}$$

and there exists a constant P_0 such that the Poincaré inequality

$$\forall\, x \in M, \ \forall\, r > 0, \quad \int_{B(x,r)} |f - f_B|^2 d\mu \le P_0 r^2 \int_{B(x,2r)} |\nabla f|^2 d\mu \tag{5.4.10}$$

is satisfied.

Theorem 5.4.12 *Assume that (M, g) is a complete Riemannian manifold such that (5.4.9), (5.4.10) are satisfied. Then the heat kernel $h(t, x, y)$ satisfies the two-sided Gaussian bound*

$$\frac{c_1 \exp\left(-C_1 d(x, y)^2/t\right)}{V(x, \sqrt{t})} \le h(t, x, y) \le \frac{C_2 \exp\left(-c_2 d(x, y)^2/t\right)}{V(x, \sqrt{t})}.$$

Moreover, for any integer k,

$$|\partial_t^k h(t, x, y)| \le \frac{A_k \exp\left(-c_2 d(x, y)^2/t\right)}{t^k V(x, \sqrt{t})}$$

and

$$|\partial_t^k h(t, x, y) - \partial_t^k h(t, x, z)| \le A_k \left(\frac{d(y, z)}{t}\right)^{\alpha/2} \frac{\exp\left(-c_2 d(x, y)^2/t\right)}{t^k V(x, \sqrt{t})}$$

for all $x \in M$, $t > 0$ and $z, y \in M$, $d(z, y) \le \sqrt{t}$.

If the hypotheses above are relaxed so that one only assumes that (5.4.9), (5.4.10) hold for $0 < r < R$, for some fixed $R > 0$, then the same conclusions hold with the restriction that $0 < t < R^2$.

Corollary 5.4.13 *Let (M, g) be a complete manifold. Assume that (5.4.10) and (5.4.9) are satisfied. Then (M, g) admits a positive symmetric Green function if and only if $\int^\infty V(x, \sqrt{t})^{-1} dt < \infty$. Moreover, if this condition holds, the Green function $G(x, y)$ satisfies*

$$c \int_{d(x,y)^2}^\infty \frac{dt}{V(x, \sqrt{t})} \le G(x, y) \le C \int_{d(x,y)^2}^\infty \frac{dt}{V(x, \sqrt{t})}$$

and, for some positive α,

$$\frac{|G(x, y) - G(x, z)|}{d(y, z)^\alpha} \le C \int_{d(x,y)^2}^\infty \frac{dt}{t^{\alpha/2} V(x, \sqrt{t})}$$

for all $x, y, z \in M$, $x \ne y$, $d(y, z) \le d(x, y)/2$.

To see this, let us first observe that the Green function, if it exists, is the kernel of $\Delta^{-1} = \int_0^\infty e^{-t\Delta} dt$. See, e.g., [34]. Moreover, it is not hard to argue that the Green function exists if and only if the integral $\int_0^\infty h(t, x, y) dt$ is finite for all $x \neq y$ and that

$$G(x, y) = \int_0^\infty h(t, x, y) dt.$$

Now it suffices to apply the bounds of Theorem 5.4.12 and note that

$$\int_0^{d^2} h(t, x, y) dt \leq C d^2 / V(x, d)$$

and

$$\int_{d^2}^\infty h(t, x, y) dt \geq c d^2 / V(x, d)$$

when $d = d(x, y)$.

Finally, we state a more explicit bound on G under an additional assumption.

Corollary 5.4.14 *Assume that (5.4.10) and (5.4.9) are satisfied. Assume that the complete manifold (M, g) satisfies*

$$\forall\, x \in M,\ \forall\, r > 0,\ \forall\, s \in (0, r),\quad \frac{V(x, r)}{V(x, s)} \geq c \left(\frac{r}{s}\right)^2.$$

Then M admits a Green function G which satisfies

$$c \frac{d(x, y)^2}{V(x, d(x, y))} \leq G(x, y) \leq C \frac{d(x, y)^2}{V(x, d(x, y))}$$

and, for some positive α,

$$\frac{|G(x, y) - G(x, z)|}{d(y, z)^\alpha} \leq C \frac{d(x, y)^{2-\alpha}}{V(x, d(x, y))}$$

for all $x, y, z \in M$, $x \neq y$, $d(y, z) \leq d(x, y)/2$.

5.5 The parabolic Harnack principle

In the two sections above, we saw how a scale-invariant Poincaré inequality and the doubling condition imply a scale-invariant parabolic Harnack inequality. It is quite remarkable that, in fact, Poincaré inequality and doubling are equivalent to the validity of this parabolic Harnack principle. This yields a characterization of the parabolic Harnack principle as announced at the beginning of this chapter. As we will point out below,

both directions of the equivalence between "Poincaré inequality, doubling"
and "parabolic Harnack inequality" can be useful.

Let us comment on the history of parabolic Harnack inequalities. In [64],
J. Moser refers to works of Hadamard [37] and Pini [70] from 1954 concerning the case of constant coefficients in $\mathbb{R}^n$. At the end of his celebrated
1958 paper [67], J. Nash states a parabolic Harnack inequality for positive
solutions of parabolic uniformly elliptic second order differential equations
in divergence form. His approach is to derive the Harnack inequality from
bounds on the fundamental solution. However, Nash's statement is incorrect. A complete (and correct) implementation of Nash's ideas in this
direction was later given in [23]. In 1961, Moser published his famous iterative argument [63], giving a proof of the elliptic Harnack inequality for
positive solutions of uniformly elliptic equations. In [64], published in 1964,
Moser adapts his own iterative argument to the case of parabolic equations.
An interesting account is given in [71]. Moser's iteration has been used
and adapted in hundreds of papers. Let us mention in particular the works
of Aronson, Aronson and Serrin, and Trudinger. Some references are in
[4, 30, 71, 75, 76].

One of the reasons for the success of Moser's technique is that it only
depends on a small number of functional inequalities: essentially, Poincaré
and Sobolev inequalities. It can thus be used in many different situations.
Another very important feature of Moser's iteration is that it is a local
technique. This is what makes it most useful in the Riemannian context
as illustrated in the previous sections. To understand this fully, the reader
should compare it with Nash's ideas as developed in [7, 23]. Nash's ideas
require global hypotheses to be implemented successfully. This point was
not completely understood until recently. In fact, the use of Moser's iteration in the context of Riemannian geometry has often been restricted by
the incorrect belief that it would require a *global Sobolev inequality* to yield
global results. See for instance [88, page 202] and the use of Moser's technique in [14, 15] and [89]. See also the introduction of [32]. The observation
that a suitable family of local Sobolev inequalities yields good large scale
results in a Riemannian context appears in [75]. This idea however is im-
plicit in several papers from the 1980s concerning subelliptic operators. See
[47, 25] for pointers to this literature. In the subelliptic context, only small
scales are usually considered but the point of a good localization technique
as developed above is to make all scales, small or large, look alike. It is
interesting to observe how this simple idea enhances some of the techniques
developed in [14].

A crucial step towards a better understanding of the geometric meaning
of parabolic Harnack inequalities was made independently by A. Grigor'yan
[32] and the author [74] (Grigor'yan's work was done and published earlier):
it is Poincaré inequality, not Sobolev inequality that is crucial for a full
parabolic Harnack inequality to holds. This is not so obvious from previous

works on the subject which tend to emphasize the role of Sobolev inequality. See for instance the discussion of Sobolev and Poincaré inequalities as well as the discussion concerning Moser's iteration in [89]. The papers [32, 74] each contain a different proof of the fact that a scale-invariant parabolic Harnack principle holds as soon as a scale-invariant Poincaré inequality and the doubling volume condition are satisfied. The proof in [74] is based on Moser's iteration. It amounts to proving that Poincaré inequality and doubling imply a family of local Sobolev inequalities. See Theorem 5.3.3. The approach taken in [32] is different and more original, avoiding the explicit use of any form of Sobolev inequality altogether. See the introduction of [32].

What makes the contribution of [32, 74] remarkable is that the scale-invariant Poincaré inequality and the doubling volume condition are not only *sufficient* but also *necessary* for a scale-invariant parabolic Harnack principle to hold (details are given in Section 5.5.1 below). Both [32] and [74] noticed that the doubling condition is necessary. As observed in [74], the scale-invariant Poincaré inequality is also a necessary condition, thanks to an argument due to Kusuoka and Stroock [50].

5.5.1 Poincaré, doubling, and Harnack

The next theorem is one of the major results presented in this monograph. Half of it has already been proved in the two previous sections.

Theorem 5.5.1 *Fix $0 < R \le \infty$ and consider the following properties:*

(i) *There exists P_0 such that, for any ball $B = B(x, r)$, $x \in M$, $0 < r < R$, and for all $f \in C^\infty(B)$,*

$$\int_B |f - f_B|^2 d\mu \le P_0 \, r^2 \int_B |\nabla f|^2 d\mu.$$

(ii) *There exists D_0 such that, for any ball $B = B(x, r)$, $x \in M$, $0 < r < R$,*

$$\mu(2B) \le D_0\mu(B).$$

(iii) *There exists a constant A such that, for any ball $B = B(x, r)$, $x \in M$, $0 < r < R$ and for any smooth positive solution u of $(\partial_t + \Delta)u = 0$ in the cylinder $(s - r^2, s) \times B(x, r)$, we have*

$$\sup_{Q_-}\{u\} \le A \inf_{Q_+}\{u\} \tag{5.5.1}$$

with

$$
\begin{aligned}
Q_- &= (s - (3/4)r^2, s - (1/2)r^2) \times B(x, (1/2)r)\\
Q_+ &= (s - (1/4)r^2, s) \times B(x, (1/2)r).
\end{aligned}
$$

Then the conjunction of (i) and (ii) is equivalent to (iii).

Although it is not strictly necessary to use this fact in the proof given below, it is often helpful to remember that $t \mapsto h(t, x, x)$ is non-increasing.

That (i) and (ii) imply (iii) is the content of Theorem 5.4.3. That (iii) implies (ii) can be seen as follows. The proof of Theorem 5.4.10 shows that (iii) implies the on-diagonal heat kernel lower bound (note that one cannot use (ii) here)

$$\forall\, x \in M, \quad \forall\, t \in (0, R^2), \quad h(t, x, x) \geq aV(x, 2\sqrt{t})^{-1}. \tag{5.5.2}$$

Inequality (5.5.2) can be complemented with a matching upper bound as follows. Applying (iii), for all $x, y \in M, t \in (0, R^2)$, with $d(x, y) < (1/2)\sqrt{t}$, we have

$$h(t, x, x) \leq Ah(2t, x, y).$$

Integrating over $B(x, (1/2)\sqrt{t})$ yields

$$V(x, (1/2)\sqrt{t})h(t, x, x) \leq A \int h(2t, x, y)d\mu(y) \leq A.$$

Thus

$$\forall\, x \in M, \quad \forall\, t \in (0, R^2), \quad h(t, x, x) \leq AV(x, (1/2)\sqrt{t})^{-1}. \tag{5.5.3}$$

Finally, a last application of (iii) yields that, for any fixed $k \geq 1$,

$$\forall\, x \in M, \quad \forall\, t \in (0, R^2), \quad h(t, x, x) \leq A_k h(kt, x, x).$$

This, together with (5.5.2), (5.5.3), shows that the doubling property (ii) holds true.

Let us prove now that (iii) implies (i). For this we need to introduce the Laplacian with Neumann boundary condition on any given metric ball $B \subset M$. However, metric balls do not necessarily have a smooth boundary so it is better to define this operator without explicit reference to a boundary condition. We can proceed as follows. Fix a ball $B \subset M$. Consider the subspace $\mathcal{D}^\infty \subset \mathcal{C}^\infty(B)$ of those smooth functions f such that

$$\forall\, g \in \mathcal{C}^\infty(B), \quad \int_B g\Delta f d\mu = \int_B \nabla g \cdot \nabla f d\mu.$$

Observe that $\mathcal{D}^\infty$ contains $\mathcal{C}_0^\infty(B)$ and thus is dense in $L^2(B)$. Observe also that the operator Δ with dense domain $\mathcal{D}^\infty$ is symmetric, i.e.,

$$\forall\, f, g \in \mathcal{D}^\infty, \quad \int_B g\Delta f d\mu = \int_B f\Delta g d\mu.$$

Also, $Q_B^N(f, f) = \int_B f\Delta f d\mu$ is non-negative for all $f \in \mathcal{D}^\infty$. It follows that the quadratic form $(Q_B^N, \mathcal{D}^\infty)$ is closable and its minimal closure $(Q_B^N, \mathcal{D})$

is associated with a self-adjoint extension of Δ which we denote by Δ_B^N. Fortunately, we will not need to understand better what the mysterious domains $\mathcal{D}^\infty$, $\mathcal{D}$ are. If B has a smooth boundary, then

$$\forall f, g \in \mathcal{C}^\infty(B), \quad \int_B g\Delta f d\mu = \int_B \nabla g \cdot \nabla f d\mu + \int_{\partial B} g\partial_\nu f d\mu_{n-1}$$

where ν is the exterior normal along ∂B. It then follows from the above construction that any function f in $\mathcal{D}^\infty$ must satisfy $\partial_\nu f = 0$, that is, f satisfies the Neumann boundary condition.

The closed form $(Q_B^N, \mathcal{D})$ on $L^2(B)$ is a Dirichlet form and the associated semigroup $H_t^{B,N} = e^{-t\Delta_B^N}$ is a self-adjoint Markov semigroup on $L^2(B)$. Observe that constant functions are indeed in $\mathcal{D}$ so that clearly $H_t^{B,N}1 = 1$. For any function $f \in \mathcal{C}^\infty(B)$, $u(t,x) = H_t^{B,N}f(x)$ is a solution of the heat equation $(\partial_t + \Delta)u = 0$ in $(0,\infty) \times B$. In particular, $H_t^{B,N}$ admits a smooth kernel

$$(t,x,y) \mapsto h_B^N(t,x,y), \quad (t,x,y) \in (0,\infty) \times B \times B.$$

Note however that there is no known method, in general, to give a uniform bound (upper or lower) on this kernel for a fixed t and all $(x,y) \in B \times B$. This would require some analysis of the boundary of B. We say that h_B^N is the Neumann heat kernel in the ball B.

Theorem 5.5.2 *Fix $0 < R < \infty$ and assume that condition (iii) of Theorem 5.5.1 holds true. Then the Neumann heat kernel in any ball B of radius $0 < r < R$ satisfies*

$$\frac{a}{V(y,\sqrt{t})} \leq h^{B,N}(t,y,z) \leq \frac{A}{V(y,\sqrt{t})}$$

for all $t \in (0,r^2)$, $y, z \in B(x,r/2)$ with $z \in B(y,\sqrt{t})$.

To prove Theorem 5.5.2, it is handy (although not strictly necessary) to use here the fact that (iii) implies (ii) (this has been proved above). Note first that

$$\forall t \in (0,r^2), \quad \forall y \in B(x,r/2), \quad \frac{a}{V(y,\sqrt{t})} \leq h^{B,N}(t,y,y) \leq \frac{A}{V(y,\sqrt{t})}.$$

This can be proved by the argument used for (5.5.2), (5.5.3). Once this has been established, (iii) easily yields the two sided inequality of Theorem 5.5.2. What we really need from Theorem 5.5.2 is the lower bound

$$h_B^N(r^2,y,z) \geq aV^{-1}, \quad y, z \in (1/2)B \tag{5.5.4}$$

for any $0 < r < R$ and any ball $B(x,r) = B$ with $V = V(x,r) = \mu(B)$. With this lower bound at hand, for $y \in (1/2)B$ and $f \in \mathcal{C}^\infty(B)$, write

$$H_{r^2}^{B,N}[f - H_{r^2}^{B,N}f(y)]^2(y) = \int_B h_B^N(r^2,y,z)|f(z) - H_{r^2}^{B,N}(y)|^2 d\mu(z)$$

$$\geq\ aV^{-1}\int_{(1/2)B}|f(z)-H_{r^2}^{B,N}(y)|^2d\mu(z)$$

$$\geq\ aV^{-1}\int_{(1/2)B}|f-f_{(1/2)B}|^2d\mu$$

where $f_{(1/2)B}$ is the mean of f over the ball $(1/2)B$. The last inequality follows from the well-known fact that the mean of f over Ω realizes the minimum of $c\mapsto\int_\Omega|f-c|^2d\mu$ over all reals c, when the integral is over a bounded domain Ω (here $\Omega=(1/2)B$).

Integrating over the ball $(1/2)B$, we get

$$\int_B H_{r^2}^{B,N}[f-H_{r^2}^{B,N}f(y)]^2(y)d\mu(y) \tag{5.5.5}$$

$$\geq\ \int_{(1/2)B}H_{r^2}^{B,N}[f-H_{r^2}^{B,N}f(y)]^2(y)d\mu(y)$$

$$\geq\ a\int_{(1/2)B}|f-f_{(1/2)B}|^2d\mu. \tag{5.5.6}$$

But a simple computation shows that

$$\int_B H_{r^2}^{B,N}[f-H_{r^2}^{B,N}f(y)]^2(y)d\mu(y)=\|f\|_{B,2}^2-\|H_{r^2}^{B,N}f\|_2^2 \tag{5.5.7}$$

and we have

$$\begin{aligned}
\|f\|_{B,2}^2-\|H_{r^2}^{B,N}f\|_2^2 &=\ -\int_0^{r^2}\partial_s\|H_s^{B,N}f\|_2^2ds\\
&=\ 2\int_0^{r^2}\langle\Delta_B^N H_s^{B,N}f,H_s^{B,N}f\rangle ds\\
&=\ 2\int_0^{r^2}Q_B^N(H_s^{B,N}f,H_s^{B,N}f)ds\\
&\leq\ 2r^2Q_B^N(f,f)=2r^2\int_B|\nabla f|^2d\mu. \tag{5.5.8}
\end{aligned}$$

To see the last inequality, observe that $s\mapsto Q_B^N(H_s^{B,N}f,H_s^{B,N}f)$ is a non-increasing function. This can be proved by noting that

$$Q_B^N(H_s^{B,N}f,H_s^{B,N}f)=\langle\Delta_B^N H_s^{B,N}f,H_s^{B,N}f\rangle=\|(\Delta_B^N)^{1/2}H_s^{B,N}f\|_2^2.$$

It follows from (5.5.6), (5.5.7) and (5.5.8) that

$$\int_{(1/2)B}|f-f_{(1/2)B}|^2d\mu\leq 2a^{-1}r^2\int_B|\nabla f|^2d\mu.$$

This proves that (iii) implies (i) (see Lemma 5.3.1).

Let us mention the following result, which complements Theorem 5.5.1.

Theorem 5.5.3 *Fix $0 < R \leq \infty$. Let M be a complete manifold. The heat kernel $h(t,x,y)$ satisfies the two-sided Gaussian inequality*

$$\frac{c_1 \exp\left(-d(x,y)^2/C_1 t\right)}{V(x,\sqrt{t})} \leq h(t,x,y) \leq \frac{C_2 \exp\left(-d(x,y)^2/c_2 t\right)}{V(x,\sqrt{t})}$$

for all $x, y \in M$ and $t \in (0, R)$ if and only if M satisfies the conditions (i) and (ii) of Theorem 5.5.1 for the same R.

We only outline the proof in the case $R = \infty$ (the case $0 < R < \infty$ is similar). First, one shows that the Gaussian lower bound implies doubling. Indeed, integrating over the ball of radius $2r$ with $r = \sqrt{t}$ yields

$$\frac{c_1 e^{-4C_1}}{V(x,2r)} V(x,r) \leq 1$$

because $\int_N h(t,x,y)dy \leq 1$. With this observation, one can use the two-sided Gaussian bound to obtain a lower bound on the Dirichlet heat kernel on any ball B. This lower bound is of the form

$$\inf_{x,y \in \epsilon B} h_B^D(t,x,y) \geq \frac{c}{\mu(B)}$$

for all $0 < t < \alpha r(B)^2$, with ϵ, α positive but small enough. It follows that the Neumann heat kernel h_B^N (which is always larger than h_B^D) satisfies the same lower bound and this yields the desired Poincaré inequality. Details can be found in [77, Proposition 2].

5.5.2 Stochastic completeness

Let us first explain the title of this section. Associated to the Laplace–Beltrami operator Δ on M is a stochastic process called Brownian motion with the property that the heat kernel $h(t,x,y)$ describes the probability of reaching y at time t starting from x. More precisely, the probability of reaching a neighborhood U of y at time t, starting from x, is equal to $\int_U h(t,x,y)dy$. One says that M is stochastically complete if this process stays on M up to any finite time. That is, M is stochastically complete if and only if

$$\int_M h(t,x,y)dy = 1.$$

It is not hard to see that this true for one $(x,t) > 0$ if and only if it is true for all $(x,t) > 0$. If this fails, it means that Brownian motion escapes to infinity in finite time. The question of stochastic completeness for Brownian motion on complete manifolds has been studied by many people, including Gaffney [27]. An early and very satisfactory criterion in terms of volume growth was obtained by A. Grigor'yan in [31]. See [34] for a thorough review of the problem. The main purpose of this section is to obtain the following result, needed in the next section.

Theorem 5.5.4 *Fix $R > 0$. Assume that the complete Riemannian manifold M satisfies the doubling condition*

$$\forall r \in (0, R), \quad V(x, 2r) \leq D_0 V(x, r).$$

Then M is stochastically complete, that is,

$$\forall x \in M, \ \forall t > 0, \ \int_M h(t, x, y) dy = 1.$$

This will be an easy consequence of the following unicity result, which is of independent interest.

Theorem 5.5.5 *Let M be a complete Riemannian manifold. Fix $T > 0$. Let u be a solution of $(\partial_t + \Delta) u = 0$ in $M_T = (0, T) \times M$ with initial condition $u(0, \cdot) \equiv 0$. Fix $o \in M$ and suppose that there exists C such that for any ball $B(o, r)$, $r > 0$,*

$$\int_0^T \int_{B(o,r)} |u(s, x)|^2 dx ds \leq e^{C(1+r)^2}.$$

Then $u = 0$ in M_T.

Let us first check that Theorem 5.5.4 follows from Theorem 5.5.5. Set $v(t, x) = \int_M h(t, x, y) dy$ and $u = 1 - v$. Then $0 \leq u \leq 1$ and $u(0, x) \equiv 0$. Thus for any ball $B(o, r)$, $r > 0$,

$$\int_0^T \int_{B(o,r)} |u(s, x)|^2 dx ds \leq V(o, r).$$

By Lemma 5.2.7, the hypothesis of Theorem 5.5.4 implies the volume upper bound

$$V(o, r) \leq e^{C(1+r)}.$$

Thus Theorem 5.5.5 applies to u and shows that u must be constant equal to 0 in M_T, for any finite T. That is, $v(t, x) = \int_M h(t, x, y) dy = 1$ for all (t, x) as desired.

We now prove Theorem 5.5.5. The proof is taken from [31] but this line of argument is well known. See, e.g., [2]. Let ρ be a function such that $|\nabla \rho| \leq 1$, fix $s > 0$ and set

$$g(t, x) = \frac{\rho(x)^2}{4(t - s)}, \quad t \neq s.$$

From this definition, it follows that

$$\partial_t g + |\nabla g|^2 \leq 0 \tag{5.5.9}$$

Let η be a smooth function with compact support. Then, as u is a solution of $(\partial_t + \Delta)u = 0$,

$$\int_{\tau-a}^{\tau} \int_M (\partial_t u) u \, \eta^2 \, e^g \, d\mu dt = -\int_{\tau-a}^{\tau} \int_M (\Delta u) \, u \, \eta^2 \, e^g \, d\mu dt$$

which we can rewrite as

$$\int_M u^2 \eta^2 e^g d\mu \Big|_{\tau-a}^{\tau} - \int_{\tau-a}^{\tau} \int_M |u|^2 \eta^2 e^g \partial_t g d\mu dt = -2 \int_{\tau-a}^{\tau} \int_M (\Delta u) u \eta^2 e^g d\mu dt.$$

Integrating the right-hand side by parts yields

$$\int_M (\Delta u) u \eta^2 e^g d\mu = \int_M |\nabla u|^2 \eta^2 e^g d\mu + \int_M \nabla u \cdot (2e^g \eta \nabla \eta + \eta^2 e^g \nabla g) \, d\mu$$

$$\geq \int_M |\nabla u|^2 \eta^2 e^g d\mu$$

$$-\frac{1}{2} \int_M |\nabla u|^2 \eta^2 e^g d\mu - 2 \int_M |u|^2 |\nabla \eta|^2 e^g d\mu$$

$$-\frac{1}{2} \int_M |\nabla u|^2 \eta^2 e^g d\mu - \frac{1}{2} \int_M |u|^2 |\nabla g|^2 \eta^2 e^g d\mu$$

$$\geq -2 \int_M |u|^2 |\nabla \eta|^2 e^g d\mu - \frac{1}{2} \int_M |u|^2 |\nabla g|^2 \eta^2 e^g d\mu.$$

Thus

$$\int_M u^2 \eta^2 e^g d\mu \Big|_{\tau-a}^{\tau} - \int_{\tau-a}^{\tau} \int_M |u|^2 \eta^2 e^g \partial_t g d\mu dt$$

$$\leq 4 \int_{\tau-a}^{\tau} \int_M |u|^2 |\nabla \eta|^2 e^g d\mu dt + \int_{\tau-a}^{\tau} \int_M |u|^2 |\nabla g|^2 \eta^2 e^g d\mu dt.$$

By (5.5.9), this gives

$$\int_M u^2 \eta^2 e^g d\mu \Big|_{\tau-a}^{\tau} \leq 4 \int_{\tau-a}^{\tau} \int_M |u|^2 |\nabla \eta|^2 e^g d\mu dt.$$

We now fix $r \geq 1$ and choose η, with support in the ball $B(o, 4r)$, equal to 1 in $B(o, 2r)$ and such that $0 \leq \eta \leq 1$, $|\nabla \eta| \leq 1/r$. This yields

$$\int_{B(o,r)} u^2(\tau, x) e^{g(\tau,x)} dx \leq \int_{B(o,4r)} u^2(\tau - a, x) e^{g(\tau-a,x)} dx$$

$$+\frac{4}{r^2} \int_{\tau-a}^{\tau} \int_{B(o,4r)\backslash B(o,2r)} |u|^2 e^g d\mu dt.$$

We also choose

$$\rho(x) = \inf\{d(x, z) : z \in B(o, r)\}, \quad s = \tau + a$$

so that

$$g(t,x) = -\frac{\rho(x)^2}{\tau + a - t} \leq 0$$

for $t \leq \tau$ and $g(t,x) = 0$ if $x \in B(o,r)$. It follows that the factor e^g can be dropped in the two first integrals. Moreover, for $x \in B(o,4r) \setminus B(o,2r)$ and $\tau - a \leq t \leq \tau$,

$$g(t,x) = -\frac{\rho(x)^2}{\tau + a - t} \leq -\frac{r^2}{2a}.$$

Hence

$$\int_{B(o,r)} u^2(\tau,x)dx \leq \int_{B(o,4r)} u^2(\tau - a, x)dx$$
$$+ \frac{4e^{-r^2/2a}}{r^2} \int_{\tau-a}^{\tau} \int_{B(o,4r)} |u(t,x)|^2 dx dt.$$

By the hypothesis made on u in Theorem 5.5.5 we can now choose a small enough so that

$$e^{-r^2/2a} \int_{\tau-a}^{\tau} \int_{B(o,4r)} |u|^2 d\mu dt \leq e^{-(r^2/2a)+C(1+r)^2} \leq 1/4.$$

For this choice of a, we obtain

$$\int_{B(o,r)} u^2(\tau,x)dx \leq \int_{B(o,4r)} u^2(\tau - a, x)dx + \frac{1}{r^2}.$$

Finally, fix $0 < s < T$, $R > 0$. Pick m large enough and apply the above m times with $r = 4^k R$, $a = s/m$, $\tau_k - \tau_{k-1} = a$, $\tau_0 = 0$, $k = 1, \ldots, m$. This yields

$$\int_{B(o,R)} u^2(s,x)dx \leq \int_{B(o,4^m R)} u^2(0,x)dx + \frac{1}{R^2}\sum_{1}^{m} 4^{-2k} \leq \frac{1}{R^2}.$$

Since this holds for all R, it follows that $u(s,\cdot) \equiv 0$ as desired.

5.5.3   Local Sobolev inequalities and the heat equation

The aim of this section is to clarify the role of local Sobolev inequalities in the study of the heat equation presented in this chapter. The interested reader should compare the results below with those contained in [33], which are based on a different but equivalent set of functional inequalities called Faber–Krahn inequalities. Also, the results below should be compared with Theorem 5.5.1.

Fix $0 < R \leq \infty$, $\nu > 2$, and consider the following properties:

- For any ball B of radius $0 < r(B) < R$ and for any $f \in \mathcal{C}_0^\infty(B)$,

$$\left(\int |f|^{2\nu/(\nu-2)} d\mu \right)^{(\nu-2)/\nu} \leq C_1 \frac{r(B)^2}{\mu(B)^{2/\nu}} \int \left(|\nabla f|^2 + r(B)^{-2} |f|^2 \right) d\mu \tag{5.5.10}$$

- For any concentric balls B, B' of radius $0 < r < r' < R$,

$$\mu(B') \leq C_2 (r'/r)^\nu \mu(B). \tag{5.5.11}$$

- For any $0 < t < R^2$, any ball B of radius $\sqrt{t}$ and any positive solution u of $(\partial_t + \Delta)u = 0$ in $Q = (t/2, t) \times B$,

$$\sup_{Q_+} \{u^2\} \leq \frac{C_3}{t\mu(B)} \int_Q u^2 d\mu \tag{5.5.12}$$

where $Q_+ = (t/2, t) \times (1/2)B$.

- For any $x \in M$ and $0 < t < R^2$,

$$h(t, x, x) \leq \frac{C_4}{V(x, \sqrt{t})}. \tag{5.5.13}$$

- For any $x, y \in M$ and any $0 < t < R^2$,

$$h(t, x, y) \leq C_5 V(x, \sqrt{t})^{-1} \exp\left(-c \frac{d(x,y)^2}{t} \right). \tag{5.5.14}$$

- There exists $\epsilon > 0$ such that, for any $0 < t < \epsilon R^2$ and any $x \in M$,

$$h(t, x, x) \geq \frac{C_6}{V(x, \sqrt{t})}. \tag{5.5.15}$$

Theorem 5.5.6 *Let M be a complete non-compact manifold.*

(1) The local Sobolev inequality (5.5.10) implies the inequalities (5.5.11), (5.5.12), (5.5.13), (5.5.14).

(2) Assume that (5.5.11) holds true. Then the local Sobolev inequality (5.5.10), the mean value inequality (5.5.12), the on-diagonal heat kernel upper bound (5.5.13), and the Gaussian heat kernel upper bound (5.5.14) are equivalent properties.

(3) Assume again that (5.5.11) holds true. Then the local Sobolev inequality (5.5.10) (or any of the equivalent properties listed in 2 above) implies the on-diagonal lower bound (5.5.15).

The proof of the first assertion is contained in Section 5.2.1, Theorem 5.2.2, Theorem 5.2.9 and equation (5.2.17).

Let us prove the second assertion. Clearly (5.5.14) implies (5.5.13). Let us show that (5.5.12) implies (5.5.13). Indeed, applying (5.5.12) to the positive solution $u(t, y) = h(t, x, y)$ in the cylinder $Q = (t/2, t) \times B$ where B is the ball of radius $\sqrt{t}$ around x, we get

$$
\begin{aligned}
h(t, x, x)^2 &\leq \frac{C_3}{tV(x, \sqrt{t})} \int_Q h(s, x, y)^2 dy ds \\
&\leq \frac{C_3}{tV(x, \sqrt{t})} \int_{t/2}^t \int_M h(s, x, y)^2 dy ds \\
&\leq \frac{C_3}{tV(x, \sqrt{t})} \int_{t/2}^t h(2s, x, x) ds \\
&\leq \frac{C_3}{2V(x, \sqrt{t})} h(t, x, x)
\end{aligned}
$$

where we have used successively the facts that

$$
\int_M h(s, x, y)^2 dy = \int_N h(s, x, y) h(s, y, x) dy = h(2s, x, x)
$$

and that $s \mapsto h(s, x, x)$ is a non-increasing function. Simplifying by $h(t, x, x)$ we get (5.5.13) as desired.

Now it suffices to show that (5.5.13) implies the local Sobolev inequality (5.5.10). In doing so, we will use our hypothesis that (5.5.11) holds true. Fix a ball B of center x and radius $r < R$. By (5.5.11) and (5.5.13), we have

$$
h(t, x, x) \leq \frac{C_4 r(B)^\nu}{\mu(B)} t^{-\nu/2}
$$

for all $0 < t < r(B)^2$. Let us introduce the Dirichlet semigroup $H_t^{B,D}$ which is associated with the minimal closure of the form $\int_M |\nabla f|^2 d\mu$ with domain $\mathcal{C}_0^\infty(B)$ in $L^2(B, d\mu)$. Let $h_B^D(t, x, y)$ be the corresponding Dirichlet heat kernel. We need to use a classical fact: the Dirichlet heat kernel is always bounded above by the full heat kernel $h(t, x, y)$ (this follows from the maximum principle: $h - h_B^D$ is a solution of the heat equation in $(0, \infty) \times B$ and is positive on $(0, \infty) \times \partial B$ because, on the boundary, $h > 0$ and $h_B^D = 0$). We will use this fact to prove the following lemma.

Lemma 5.5.7 *Assume that* (5.5.11) *and* (5.5.13) *hold true for some* $R > 0$. *Then there exists a constant* C *such that, for any ball* B *of radius* $r(B)$ *less than* R, *the Sobolev inequality*

$$
\forall f \in \mathcal{C}_0^\infty(B), \quad \|f\|_{2\nu/(\nu-2)}^2 \leq C \frac{r(B)^2}{\mu(B)^{2/\nu}} \int \left(|\nabla f|^2 + r(B)^{-2} |f|^2 \right) d\mu
$$

is satisfied.

Setting $r = r(B)$, the hypothesis yields the estimate

$$\forall\, y \in B, \ \ e^{-t/r^2} h_B^D(t,y,y) \le e^{-t/r^2} h(t,y,y) \le C\mu(B)^{-1} r^\nu t^{-\nu/2}.$$

Applying Theorem 4.1.2 to the semigroup $e^{-t/r^2} H_t^{B,D}$, we obtain the Nash inequality

$$\|f\|_2^{2(1+2/\nu)} \le C\mu(B)^{-2/\nu} r^2 \left(\int (|\nabla f|^2 + r^{-2}|f|^2)d\mu \right) \|f\|_1^{4/\nu}.$$

The desired result then follows from the equivalence between the (local) Nash inequality and the (local) Sobolev inequality. See Corollary 3.2.12 and Section 3.2.7. This finishes the proof of the second assertion of Theorem 5.5.6.

Finally, the last assertion of Theorem 5.5.6 follows from Theorem 5.2.14 and this finishes the proof of Theorem 5.5.6.

For completeness, we note the following result. The Dirichlet heat kernel admits a discrete spectral decomposition

$$h_B^D(t,x,y) = \sum_1^\infty e^{-t\lambda_i} \phi_i(x)\phi_i(y)$$

where $\lambda_1 < \lambda_2 \le \cdots$ are the eigenvalues in non-increasing order repeated according to multiplicity and the ϕ_i are associated real eigenfunctions normalized by $\|\phi_i\|_2 = 1$ (in $L^2(B,d\mu)$). In particular,

$$e^{-t\lambda_1} \le \int_B h_B^D(t,x,x)dx.$$

This can be used to derive an estimate of the Dirichlet eigenvalue $\lambda_1(B)$.

Lemma 5.5.8 *Assume that* (5.5.11) *and* (5.5.13) *hold true for some* $R > 0$. *Then there exists* $\epsilon > 0$ *such that, for any ball* B *of radius smaller than* ϵR, *the Dirichlet eigenvalue* λ_1 *is bounded below by* $\lambda_1 \ge c/r^2$. *Here* $\epsilon > 0$ *depends on the constants appearing in the two hypotheses* (5.5.11) *and* (5.5.13).

Fix the ball B of radius $r < \epsilon R$. Let B' be a concentric ball of radius $s \in (r,R)$ to be chosen later. We have

$$e^{-s^2\lambda_1} \le \int_B h_B^D(s^2,z,z)dz$$

and

$$\forall\, z \in B', \ \ h_B^D(s^2,z,z) \le h(s^2,z,z) \le \frac{C_4}{\mu(B')}.$$

It follows that

$$e^{-s^2\lambda_1} \le C_4 \frac{\mu(B)}{\mu(B')}.$$

By Lemma 5.2.8, there exists $\gamma > 0$ such that

$$\frac{\mu(B)}{\mu(B')} \leq C(r/s)^\gamma.$$

Choosing $s = r/\epsilon$ (which indeed belongs to (r, R)), we get

$$e^{-s^2 \lambda_1} \leq C' \epsilon^\gamma.$$

If $\epsilon > 0$ is fixed small enough so that $C'\epsilon^\gamma \leq e^{-1}$ the above inequality implies that

$$\lambda_1 \geq 1/s^2 = \epsilon^2/r^2$$

as desired.

One cannot dispense with introducing a small $\epsilon > 0$ in the statement of Theorem 5.5.8. To see this, consider the connected sum of $\mathbb{R}^3$ with a unit 3-dimensional sphere. In fact, consider a family of such examples, depending on a small parameter $\eta \geq 0$ which describes the width of the smooth collar between the sphere and the flat space. When $\eta = 0$, the sphere and $\mathbb{R}^3$ are disconnected. It is possible to prove that, uniformly in η, these manifolds satisfy (5.5.11) and (5.5.13) for any finite fixed R, for instance $R = 2\pi$. This is obvious for (5.5.11). It is less obvious for (5.5.13), but it should be intuitively clear since the result is certainly true when $\eta = 0$. Now, it is not possible to lower bound uniformly the Dirichlet eigenvalues of all balls of radius 2π. Again, this should be intuitively clear because, when $\eta = 0$, any ball of radius 2π centered on the sphere is just the sphere itself and the "Dirichlet" eigenvalue vanishes (there is no boundary).

5.5.4   Selected applications of Theorem 5.5.1.

We present two related applications The first is that the parabolic Harnack principle is stable under quasi-isometries. Let M be a manifold and $g, \widetilde{g}$ be two Riemannian metrics on M. One says that these Riemannian metrics are quasi-isometric if there exists $c > 0$ such that

$$\forall x \in M, \ \ \forall X \in T_x, \ \ cg_x(X, X) \leq \widetilde{g}_x(X, X) \leq c^{-1} g_x(X, X). \qquad (5.5.16)$$

Theorem 5.5.9 *For any fixed R, $0 < R \leq \infty$, the parabolic Harnack inequality (5.5.1) is stable under quasi-isometry. That is, if (M, g) and $(M, \widetilde{g})$ are two complete Riemannian structures on M such that g and $\widetilde{g}$ are quasi-isometric, then (5.5.1) holds on (M, g) if and only if it holds for $(M, \widetilde{g})$.*

This immediately follows from Theorem 5.5.1. Indeed, the conditions (i) and (ii) are obviously stable under quasi-isometry since (5.5.16) implies

$$cg(\nabla f, \nabla f) \leq \widetilde{g}(\widetilde{\nabla} f, \widetilde{\nabla} f) \leq c^{-1} g(\nabla f, \nabla f)$$

and

$$c^{n/2} \leq \frac{d\mu}{d\widetilde{\mu}} \leq c^{-n/2}$$

where n is the topological dimension of M.

It is worth pointing out here that, because of the nature of the proof of Theorem 5.5.1, the above result stays valid even if the metrics $g, \widetilde{g}$ are not smooth but merely continuous or even measurable.

Next, we consider the situation where we have two complete Riemannian manifolds (M, g), $(\widetilde{M}, \widetilde{g})$ and a smooth map $\phi : \widetilde{M} \to M$ such that:

(a) ϕ is onto, i.e., $\phi(\widetilde{M}) = M$.

(b) For any $u \in \mathcal{C}^\infty(M)$, the function $v = u \circ \phi : \widetilde{M} \to \mathbb{R}$ satisfies

$$\widetilde{\Delta} v = (\Delta u) \circ \phi.$$

Theorem 5.5.10 *Assume that (M, g), $(\widetilde{M}, \widetilde{g})$ are two complete manifolds and that $\phi : \widetilde{M} \to M$ is a map satisfying* (a) *and* (b) *above. Fix $0 < R \leq \infty$. Assume that conditions* (i) *and* (ii) *of Theorem* 5.5.1 *are satisfied on $(\widetilde{M}, \widetilde{g})$, that is, assume that there exist $\widetilde{P}_0$, $\widetilde{D}_0$ such that for any ball $\widetilde{B} = \widetilde{B}(\widetilde{x}, r)$, $\widetilde{x} \in M$, $0 < r < R$,*

$$\forall f \in \mathcal{C}^\infty(\widetilde{B}), \quad \int_{\widetilde{B}} |f - f_{\widetilde{B}}|^2 d\widetilde{\mu} \leq \widetilde{P}_0 \, r^2 \int_{\widetilde{B}} |\widetilde{\nabla} f|^2 d\widetilde{\mu}$$

and

$$\widetilde{\mu}(2\widetilde{B}) \leq \widetilde{D}_0 \widetilde{\mu}(\widetilde{B}).$$

Then (M, g) also satisfies these two properties.

Stated as it is, this theorem is rather non-trivial. However, it will easily follow from Theorem 5.5.1. To see this, we need to show that for any $\widetilde{g}$-metric ball $\widetilde{B} = \widetilde{B}(\widetilde{x}, r) \subset \widetilde{M}$,

$$\phi(\widetilde{B}) = B(\phi(\widetilde{x}), r). \tag{5.5.17}$$

Assuming that (5.5.17) is true, let u be a positive solution of $(\partial_t + \Delta)u = 0$ in $Q = (s - r^2, s) \times B(x, r)$, $x \in M$, $s \in \mathbb{R}$, $0 < r < R$. Set $v = u \circ \phi$. Then, by (b), v is a solution of $(\partial_t + \widetilde{\Delta})v = 0$ in the set $(s - r^2, s) \times \phi^{-1}(B(x, r))$. Fix some $\widetilde{x} \in \widetilde{M}$ such that $\phi(\widetilde{x}) = x$ and set $\widetilde{B} = \widetilde{B}(\widetilde{x}, r)$, $\widetilde{Q} = (s - r^2, s) \times \widetilde{B}$. By (5.5.17), $\widetilde{Q} \subset U$. Since (i) and (ii) of Theorem 5.5.1 hold on $(\widetilde{M}, \widetilde{g})$, (iii) also holds true there. This gives

$$\sup_{\widetilde{Q}_-} \{v\} \leq A \inf_{\widetilde{Q}_+} \{v\}$$

with $\widetilde{Q}_- = (s - (3/4)r^2, s - (1/2)r^2) \times \widetilde{B}(\widetilde{x}, (1/2)r)$, $\widetilde{Q}_+ = (s - (1/4)r^2, s) \times \widetilde{B}(\widetilde{x}, (1/2)r)$. Using (5.5.17) again, it follows that

$$\sup_{Q_-}\{u\} \leq A \inf_{Q_+}\{u\}$$

with $Q_- = (s - (3/4)r^2, s - (1/2)r^2) \times B(x, (1/2)r)$, $Q_+ = (s - (1/4)r^2, s) \times B(x, (1/2)r)$. That is, (5.5.1) is satisfied on (M, g). Thus, the conditions (i) and (ii) of Theorem 5.5.1 are also satisfied.

Before proving (5.5.17), let us state some further properties of ϕ which follow from conditions (a),(b). Let $f, h \in \mathcal{C}^\infty(M)$. Set $\widetilde{f} = f \circ \phi$ and define $\widetilde{h}$ similarly. Then (a) and the formula

$$\Delta(fh) = f\Delta h + h\Delta f + g(\nabla f, \nabla h)$$

together with the similar formula on $(\widetilde{M}, \widetilde{g})$, imply that

$$\widetilde{g}(\widetilde{\nabla}\widetilde{f}, \widetilde{\nabla}\widetilde{h})(\widetilde{x}) = g(\nabla f, \nabla h)(x), \qquad (5.5.18)$$

for all $\widetilde{x}, x$ such that $\phi(\widetilde{x}) = x$.

Fix a local frame (e_i) on M around $x \in M$ so that the metric g has the canonical form (i.e., $g_x^{i,j} = \delta_{i,j}$) at x. For each i, we can find a function h_i such that $\partial_i h_i(x) = 1, \partial_j h_i(x) = 0$ if $j \neq i$. It then follows from (5.5.18) that

$$df \circ d\phi(\widetilde{\nabla}\widetilde{h}(\widetilde{x})) = \partial_i f(x).$$

Since this is true for all f, it follows that

$$d\phi(\widetilde{\nabla}\widetilde{h}_i(\widetilde{x})) = e_i(x).$$

That is, $d\phi$ is surjective. In other words, ϕ is a submersion. In fact, by (5.5.18), it is a Riemannian submersion. In particular, ϕ is open. Note that this shows that it is enough to assume that $(\widetilde{M}, \widetilde{g})$ is complete since the hypotheses (a),(b) above then imply that that (M, g) is also complete. See, e.g., [29, 2.109].

Let us now prove (5.5.17). In order to give a proof that would easily work in other settings, we will use as little Riemannian geometry as possible. A proof more in the spirit of Riemannian geometry is given in [29, 2.109]. Let us first show

$$\phi(\widetilde{B}(\widetilde{x}, r)) \subset B(\phi(\widetilde{x}), r). \qquad (5.5.19)$$

Recall that the distance d on M can be computed by the formula

$$d(x, y) = \sup\{f(x) - f(y) : f \in \mathcal{C}^\infty(M), \ |\nabla f| \leq 1\}. \qquad (5.5.20)$$

Of course, $\widetilde{d}$ can be computed from a similar formula. Let $f \in \mathcal{C}^\infty(M)$ and set $\widetilde{f} = f \circ \phi$. By (5.5.18), we have

$$|\widetilde{\nabla}\widetilde{f}|(x) = |\nabla f| \circ \phi(x). \qquad (5.5.21)$$

From (5.5.20) and (5.5.21) it follows that

$$d(x,y) \geq r \Rightarrow \tilde{d}(\tilde{x}, \tilde{y}) \geq r$$

for all $x, y \in M$, $\tilde{x}, \tilde{y} \in \widetilde{M}$ such that $\phi(\tilde{x}) = x$, $\phi(\tilde{y}) = y$. The desired inclusion (5.5.19) follows.

We now want to prove

$$\phi(\widetilde{B}(\tilde{x}, r)) \supset B(\phi(\tilde{x}), r). \tag{5.5.22}$$

To this end, we want to show, roughly, that paths can be lifted from M to $\widetilde{M}$ in an appropriate way. First let us note that it will be enough to work locally because of our assumption that both $\widetilde{M}$ and M are complete manifolds.

Fix $\tilde{x} \in \widetilde{M}$, $x \in M$, such that $\phi(\tilde{x}) = x$. As ϕ is open, we can find connected neighborhoods $\widetilde{U}$ of $\tilde{x}$ and U of x such that $\phi(\widetilde{U}) = U$. We can also assume that in $\widetilde{U}$ we have a moving orthonormal frame, say $(\widetilde{X}_i)_1^{\tilde{n}}$, where $\tilde{n}$ is the topological dimension of $\widetilde{M}$. We let $X_i = d\phi(\widetilde{X}_i)$, $1 \leq i \leq \tilde{n}$. Note that the X_i's span the tangent space at any point $x \in U$. Moreover, by (5.5.18) and the definition of the X_i's, for any vector field $X = \sum_i^{\tilde{n}} a_i X_i$ in U we have

$$g(X, X) = \sum_1^{\tilde{n}} a_i^2 = \tilde{g}(\widetilde{X}, \widetilde{X})$$

where $\widetilde{X} = \sum_i^{\tilde{n}} a_i \circ \phi \, \widetilde{X}_i$.

Now, let

$$\gamma : [0, s] \to M$$

be a smooth distance-minimizing curve parametrized by arc length and joining x to y, with y in some neighborhood of x contained in U. Let us write, as we may,

$$\partial_t \gamma(t) = \sum_1^{\tilde{n}} a_i(t) X_i(\gamma(t))$$

with smooth coefficients a_i. As γ is parametrized by arc length, we must have

$$\forall t \in [0, s], \quad g(\partial_t \gamma(t), \partial_t \gamma(t)) = 1 = \sum_1^{\tilde{n}} |a_i(t)|^2.$$

Now, define

$$\tilde{\gamma} : [0, s] \to \widetilde{M}$$

to be the solution of

$$\begin{cases} \partial_t \tilde{\gamma}(t) & = \sum_1^{\tilde{n}} a_i(t) \widetilde{X}_i(\tilde{\gamma}(t)) \\ \tilde{\gamma}(0) & = \tilde{x}. \end{cases}$$

By uniqueness of γ (as a solution of a Cauchy problem), we must have

$$\forall\, t \in [0, s], \quad \phi(\widetilde{\gamma}(t)) = \gamma(t).$$

In particular, $\widetilde{y} = \widetilde{\gamma}(s)$ is a point such that $d(\widetilde{x}, \widetilde{y}) \le s = d(x, y)$ and $\phi(\widetilde{y}) = y$. This shows that $\phi(\widetilde{B}(\widetilde{x}, r)) \supset B(x, r)$ for r small enough. The general case follows by the triangle inequality.

5.6   Examples

We describe below a few examples where the results developed in this chapter apply.

5.6.1   Unimodular Lie groups

Let us start with connected Lie groups. Let G be a connected Lie group and $(X_1, \ldots, X_n)$ be a basis of left invariant vector fields. See Section 3.3.3 for notation and details. Let us assume that G is unimodular so that $(X_1, \ldots, X_n)$ yields a Riemannian structure such that $\Delta = -\sum_1^n X_i^2$, $\nabla f = (X_1 f, \ldots, X_n f)$. The Riemannian measure is a Haar measure $d\mu$ on G.

Let $g, h \in G$ and let $\gamma_h : [0, t_h] \to G$ be a distance-minimizing curve from the neutral element e to h, parametrized by arc length. In Section 3.3.4, we proved that

$$|f(gh) - f(g)|^2 \le t \int_0^{t_h} |\nabla f(g\gamma_h(s))|^2 ds$$

for any smooth function f and all g, h. Let B be the ball of radius r around e. Then we have

$$\int_G \int_G |f(gh) - f(g)|^2 \mathbf{1}_B(g) \mathbf{1}_B(gh) dg dh$$

$$\le\ 2r \int_G \int_G \int_0^{t_h} |\nabla f(g\gamma_h(s))|^2 \mathbf{1}_B(g) \mathbf{1}_B(gh) ds dg dh$$

$$\le\ 2r \int_B \int_B \int_0^{t_h} |\nabla f(g\gamma_h(s))|^2 \mathbf{1}_{2B}(g\gamma_h(s)) \mathbf{1}_{2B}(h) ds dg dh$$

$$\le\ 2r \int_G \int_0^{t_h} \left(\int_G |\nabla f(g\gamma_h(s))|^2 \mathbf{1}_{2B}(g\gamma_h(s)) dg \right) \mathbf{1}_{2B}(h) ds dh$$

$$\le\ 2r \int_G \int_0^{t_h} \left(\int_{2B} |\nabla f(g)|^2 dg \right) \mathbf{1}_{2B}(h) ds dh$$

$$\le\ 2r^2 \mu(2B) \left(\int_{2B} |\nabla f(g)|^2 dg \right).$$

This computation makes sense for all $f \in C^\infty(2B)$. It follows that

$$\int_B |f - f_B|^2 d\mu \le 2r^2 \frac{\mu(2B)}{\mu(B)} \int_{2B} |\nabla f|^2 d\mu.$$

Let us note that (a variation of) the proof above shows that a similar Poincaré inequality holds in L^p, $1 \le p < \infty$.

By Lemma 5.3.1, we thus have proved the following result.

Theorem 5.6.1 *Let G be a unimodular connected Lie group equipped with a left-invariant Riemannian structure as above.*

(1) For all $0 < r \le 1$ and all balls B of radius r, the Poincaré inequality

$$\forall f \in C^\infty(B), \quad \int_B |f - f_B|^2 d\mu \le P_0 r^2 \int_B |\nabla f|^2 d\mu$$

is satisfied.

(2) If there exists D_0 such that $\mu(2B) \le D_0 \mu(B)$ for all balls B, then the Poincaré inequality

$$\forall f \in C^\infty(B), \quad \int_B |f - f_B|^2 d\mu \le P_0 r^2 \int_B |\nabla f|^2 d\mu$$

is satisfied for all balls B of radius $r > 0$.

As already mentioned at the end of Section 3.3.4, nilpotent Lie groups satisfy the doubling condition $\mu(2B) \le D_0 \mu(B)$ and thus (2) above applies to these groups. That is, any nilpotent Lie group equipped with a left-invariant Riemannian metric satisfies the doubling condition $\mu(2B) \le D_0 \mu(B)$ and the scale-invariant Poincaré inequality

$$\forall f \in C^\infty(B), \quad \int_B |f - f_B|^2 d\mu \le P_0 r^2 \int_B |\nabla f|^2 d\mu.$$

Thus, by Theorem 5.5.1, such a group also satisfies the scale-invariant parabolic Harnack principle (5.5.1) with $R = \infty$.

Let us describe in some detail the simplest such example, the three dimensional Heisenberg group $\mathbb{H}_1$. This is the group of all three by three upper-triangular matrices with 1's on the diagonal:

$$\mathbb{H}_1 = \left\{ \begin{pmatrix} 1 & x & z \\ 0 & 1 & y \\ 0 & 0 & 1 \end{pmatrix} : x, y, z \in \mathbb{R} \right\}. \tag{5.6.1}$$

Thus we can view $\mathbb{H}_1$ as $\mathbb{R}^3$ equipped with the product

$$(x, y, z) \bullet (x', y', z') = (x + x', y + y', z + z' + xy').$$

Observe that, up to the choice of a multiplicative positive constant, the Haar measure on $\mathbb{H}_1$ is just the Lebesgue measure in $\mathbb{R}^3$, which is both left- and right-invariant. If we let X, Y, Z be the left-invariant vector fields with

$$X(0) = \partial_x, \quad Y(0) = \partial_y, \quad Z(0) = \partial_z$$

then, in the $g = (x, y, z)$ coordinate system,

$$X(g) = \partial_x, \quad Y(g) = \partial_y + x\partial_z, \quad Z(g) = \partial_z.$$

We can now turn $\mathbb{H}_1$ into a Riemannian manifold so that at each point g, $(X(g), Y(g), Z(g))$ is an orthonormal basis of the tangent space. The Riemannian measure is then exactly the Lebesgue measure, the length of the gradient of a function f is given by

$$|\nabla f|^2 = |Xf|^2 + |Yf|^2 + |Zf|^2$$

and the Laplace–Beltrami operator is

$$\Delta f = -(X^2 f + Y^2 f + Z^2 f) = -(f_{xx}^2 + f_{yy}^2 + (1 + x^2)f_{zz}^2 + 2xf_{yz}).$$

We claim that the volume growth function $V(g, r) = V(r)$ satisfies

$$V(r) \approx \begin{cases} r^3 & \text{if } 0 < r \leq 1 \\ r^4 & \text{if } r \geq 1. \end{cases}$$

Observe that such volume functions have indeed the doubling property. We refer the reader to [87] for details and other references. For small r, the result follows from the fact that $\mathbb{H}_1$ is viewed here as a homogeneous Riemannian manifold so that the volume of small balls must be uniformly comparable to the Euclidean 3-dimensional volume. To understand what happens for large r, consider the broken curve γ which starts at 0, stays tangent to X, then to Y, then to $-X$, then $-Y$, each for the same lapse of time $s > 0$. This curve has length at most $4s$ in our Riemannian metric (in fact it has length $4s$). However, the end point of this curve is $(0, 0, s^2)$. So, in time s, we can add s^2 to the z coordinate! Using this observation one can show that (up to multiplicative constants) the ball of radius $r \geq 1$ contains and is contained in rectangular boxes of dimension $r \times r \times r^2$ and thus has volume comparable to r^4.

By Theorem 5.6.1, $\mathbb{H}_1$ equipped with the Riemannian structure above satisfies the Poincaré inequality

$$\forall f \in \mathcal{C}^\infty(B), \quad \int_B |f - f_B|^2 d\mu \leq P_0 r^2 \int_B |\nabla f|^2 d\mu$$

where r is the radius of the ball B and P_0 is independent of B. Thus, by Theorem 5.5.1, it also satisfies the scale-invariant parabolic Harnack principle (5.5.1) with $R = \infty$.

5.6.2   Homogeneous spaces

Let us now consider a simple but non-trivial application of Theorem 5.5.10. Consider a connected Lie group G and a closed subgroup $H \subset G$ and let M be the (left-)coset space $M = H\backslash G = \{x = Hg, \ g \in G\}$. Let p be the canonical projection. Then M is a manifold and p, of course, is surjective and smooth. Let us assume for simplicity that both G and H are unimodular. Up to a multiplicative constant, there exists a unique measure μ on M which is invariant under the right action of G on M. Given a left-invariant Riemannian structure on G, specified by a basis of left-invariant vector fields $(X_1, \ldots, X_n)$, we can define a Riemannian structure on M such that p is a Riemannian submersion (see, e.g., [29, 2.28]). It is interesting to note that one can compute the length of the gradient and the Laplace–Beltrami operator on M purely in terms of the vector fields $Y_i = dp(X_i)$. Indeed, for any smooth function f on M,

$$|\nabla f| = \sum_1^n |Y_i f|^2, \quad \Delta f = \sum_1^n Y_i^2 f.$$

Of course, setting $\tilde{f} = f \circ p$, we have

$$\Delta_G \tilde{f} = -\sum_1^n X_i^2 \tilde{f} = -\sum_1^n (Y_i^2 f) \circ p = (\Delta_M f) \circ p.$$

Thus, Theorem 5.5.10 applies in this case.

Theorem 5.6.2 *Let G be a unimodular connected Lie group equipped with a left-invariant Riemannian structure. Let H be a closed unimodular subgroup of G. Let $M = H\backslash G$ and equip M with the unique Riemannian structure such that the canonical projection p is a Riemannian submersion.*

(1) *For any fixed $R_0 > 0$, there exist P_0 and D_0 such that the Poincaré inequality*

$$\forall f \in \mathcal{C}^\infty(B), \quad \int_B |f - f_B|^2 d\mu \le P_0 r^2 \int_B |\nabla f|^2 d\mu$$

and the doubling property

$$\mu(2B) \le D_0 \mu(B)$$

are satisfied for all $0 < r \le R_0$ and all balls $B \subset M$ of radius r.

(2) *If the group G satisfies the global doubling property*

$$\forall B \subset G, \quad \mu(2B) \le D_0 \mu(B),$$

μ being the Haar measure on G, then the scale-invariant Poincaré inequality and the doubling property are satisfied uniformly for all balls in M. In particular, in this case, the Riemannian manifold $M = H\backslash G$ satisfies the parabolic Harnack principle (5.5.1) with $R = \infty$.

It is possible to give a more direct proof of this result. This is useful, for instance, in deriving the L^p versions, $1 \le p < \infty$, of the above L^2 Poincaré inequality on homogeneous spaces. See [59].

Again, let us describe in some detail a more concrete example. Let $G = \mathbb{H}_1$ be the Heisenberg group described at (5.6.1). Let H be the closed subgroup

$$H = \left\{ \begin{pmatrix} 1 & 0 & 0 \\ 0 & 1 & y \\ 0 & 0 & 1 \end{pmatrix} : y \in \mathbb{R} \right\}.$$

Then $\mathbb{H}_1/H$ can be identified with $\mathbb{R}^2 = \{(x, z) : x, z \in \mathbb{R}^2\}$ equipped with the Riemannian structure for which

$$X = \partial_x, \quad Z = \sqrt{1 + x^2} \, \partial_z$$

is an orthonormal basis at (x, z). The Laplace–Beltrami operator is

$$\Delta = -(\partial_x^2 + (1 + x^2)\partial_z^2).$$

It is a good exercise to prove by hand that the volume growth is doubling on this Riemannian manifold. Details can be found in [59]. By Theorem 5.6.2, $\mathbb{R}^2$ equipped with this Riemannian structure satisfies the Harnack principle (5.5.1) with $R = \infty$.

5.6.3   Manifolds with Ricci curvature bounded below

Following the work of S-T. Yau in the 1970s, a large number of analytic results have been obtained under lower bound hypotheses on the Ricci curvature tensor. We will not go into the details here but present some basic results regarding volume growth and Poincaré inequalities. For background on the curvature tensor, see, e.g., [12, 13, 29]. The Ricci tensor $\mathcal{R}$ of (M, g) can be considered as a symmetric two-tensor. As such, it can be compared with the metric tensor g. Manifolds with Ricci curvature bounded below are manifolds for which there exists a constant K such that

$$\mathcal{R} \ge -Kg. \tag{5.6.2}$$

That is,
$$\forall x \in M, \quad \forall X \in T_x, \quad \mathcal{R}_x(X, X) \ge -Kg_x(X, X).$$

One of the basic consequences of (5.6.2) is a control of the volume growth on M. Namely, if (5.6.2) is satisfied, the volume of a ball of radius r in M is at most that of a ball of radius r in the *model* space having the same dimension and $\mathcal{R} = -Kg$. Here, the model spaces are spaces of constant sectional curvature: spheres ($K < 0$), Euclidean spaces ($K = 0$), and hyperbolic spaces ($K > 0$). See [29, 3.101] or [13, Theorem 3.9]. In particular, one has the following estimates.

Theorem 5.6.3 *Assume that (M,g) is a complete manifold of dimension n satisfying (5.6.2) with $K \geq 0$. Then*

$$V(x,r) \leq \Omega_n r^n \exp\left(\sqrt{(n-1)K}\, r\right)$$

for all $x \in M$ and $r > 0$. Here, Ω_n is the volume of the Euclidean ball of radius 1 in $\mathbb{R}^n$.

M. Gromov observed that this result can be strengthened in a crucial way as follows.

Theorem 5.6.4 *Assume that (M,g) is a complete manifold of dimension n satisfying (5.6.2) with $K \geq 0$. Then*

$$V(x,r) \leq V(x,s)(r/s)^n \exp\left(\sqrt{(n-1)K}\, r\right)$$

for all $x \in M$ and $r \geq s > 0$. In particular, (M,g) satisfies the doubling condition

$$\mu(2B) \leq 2^n \exp\left(\sqrt{(n-1)K}\, R\right) \mu(B)$$

for each $0 < R < \infty$ and all balls B of radius $r \in (0,R)$. If $K = 0$, then $\mu(2B) \leq 2^n \mu(B)$.

As it turns out, these manifolds also satisfy scale-invariant Poincaré inequalities. This is a result of P. Buser [10]. See [13, Theorem 6.8].

Theorem 5.6.5 *Assume that (M,g) is a complete manifold of dimension n satisfying (5.6.2) with $K \geq 0$. Then for each $1 \leq p < \infty$, there exist $C_{n,p}$ and C_n such that*

$$\int_B |f - f_B|^p d\mu \leq C_{n,p}\, r^p e^{C_n\sqrt{K}\, r} \int_B |\nabla f|^p d\mu$$

for all balls $B \subset M$ of radius $0 < r < \infty$.

The proof of this result in [10] is elementary but quite subtle and intricate. We will prove an a priori slightly weaker statement which suffices to imply Theorem 5.6.5 by Corollary 5.3.5.

Theorem 5.6.6 *Assume that (M,g) is a complete manifold of dimension n satisfying (5.6.2) with $K \geq 0$. Then for each $1 \leq p < \infty$, there exist $C_{n,p}$ and C_n such that*

$$\int_B |f - f_B|^p d\mu \leq C_{n,p}\, r^p e^{C_n\sqrt{K}\, r} \int_{2B} |\nabla f|^p d\mu$$

for all balls $B \subset M$ of radius $0 < r < \infty$.

For any pair of points $(x, y) \in M \times M$, let

$$\gamma_{x,y} : [0, d(x, y)] \to M, \quad t \mapsto \gamma_{x,y}(t)$$

be a geodesic from x to y parametrized by arc length. Except for a set of $\mu \otimes \mu$ measure zero, this geodesic is unique and $\gamma_{y,x}(t) = \gamma_{x,y}(d(x, y) - t)$. Let us prove the theorem above for $p = 1$ (the same proof works for any other finite $p \geq 1$). Fix a ball B of radius r and write

$$
\begin{aligned}
\int_B |f - f_B| d\mu &\leq \frac{1}{\mu(B)} \int_B \int_B |f(x) - f(y)| dx dy \\
&\leq \frac{1}{\mu(B)} \int_B \int_B \int_0^{d(x,y)} |\nabla f(\gamma_{x,y}(s))| ds dx dy \\
&= \frac{2}{\mu(B)} \int_B \int_B \int_{d(x,y)/2}^{d(x,y)} |\nabla f(\gamma_{x,y}(s))| ds dx dy.
\end{aligned}
$$

To obtain the last equality we break the set

$$\{(x, y, s) : x, y \in B, \gamma_{x,y}(s) \in (0, d(x, y))\}$$

into two pieces

$$\{(x, y, s) : x, y \in B, \gamma_{x,y}(s) \in (d(x, y)/2, d(x, y))\}$$

and

$$\{(x, y, s) : x, y \in B, \gamma_{x,y}(s) \in (0, d(x, y)/2)\},$$

write the second piece

$$\{(x, y, s) : x, y \in B, \gamma_{y,x}(d(x, y) - s) \in (0, d(x, y)/2)\},$$

and use the $(x, y, s) \mapsto (y, x, d(x, y) - s)$ symmetry. This trick is crucial in obtaining the desired result by this method. It is taken from [49].

Now, suppose we can bound the Jacobian $J_{x,s}$ of the map

$$\Phi_{x,s} : y \mapsto \gamma_{x,y}(s)$$

from below by

$$\forall\, x, y \in B, \ \forall\, s \in [d(x, y)/2, d(x, y)], \ \ J_{x,s}(y) \geq 1/F(r) \qquad (5.6.3)$$

where r is the radius of the ball B. Then

$$
\begin{aligned}
&\int_B \int_B \int_{d(x,y)/2}^{d(x,y)} |\nabla f(\gamma_{x,y}(s))| ds dx dy \\
&\qquad = \int_B \int_B \int_{d(x,y)/2}^{d(x,y)} |\nabla f(\Phi_{x,s}(y))| ds dx dy
\end{aligned}
$$

$$\leq F(r) \int_B \int_B \int_{d(x,y)/2}^{d(x,y)} |\nabla f(\Phi_{x,s}(y))| J_{x,s}(y) ds dx dy$$

$$\leq F(r) \int_B \int_B \int_0^r |\nabla f(\Phi_{x,s}(y))| J_{x,s}(y) ds dx dy$$

$$\leq F(r) \int_0^r \left[\int_B \int_B |\nabla f(\Phi_{x,s}(y))| J_{x,s}(y) dx dy \right] ds$$

$$\leq F(r) \int_0^r \left[\int_B \left(\int_{\Phi_{x,s}(B)} |\nabla f(z)| dz \right) dx \right] ds$$

$$\leq F(r) \int_0^r \left[\int_B \int_{2B} |\nabla f(z)| dz dx \right] ds$$

$$\leq F(r) \, r \, \mu(B) \int_{2B} |\nabla f(z)| dz.$$

Hence

$$\int_B |f - f_B| d\mu \leq 2 \, r \, F(r) \int_{2B} |\nabla f(z)| dz.$$

Thus we are left with the task of proving that (5.6.3) holds with

$$F(r) \leq C_n \exp \left(C_n \sqrt{K} \, r \right) \tag{5.6.4}$$

Lemma 5.6.7 *Let M be a Riemannian manifold satisfying (5.6.2) for some $K \geq 0$. Then*

$$\forall x \in B, \ \forall s \in (d(x,y)/2, d(x,y)), \ \ J_{x,s}(y) \geq c_n \exp \left(-C_n \sqrt{K} \, r \right)$$

for all $y \in B$ not in the cut locus of x.

This is a consequence of the basic ingredient of the proof of the Bishop–Gromov Theorem [13, Theorem 3.9]. Namely, given x and y (y not in the cut locus of x) let ξ be the unit tangent vector at x such that $\partial_s \gamma_{x,y}(s)|_{s=0} = \xi$. Let $I(x, s, \xi)$ be the Jacobian of the map $(s, \xi) \mapsto \exp_x(s\xi)$. Then

$$d\mu = I(x, s, \xi) ds d\xi$$

where $d\xi$ is the usual measure on the sphere. Moreover, if we let $I_K(s)$ be the analog of $I(x, s, \xi)$ on the corresponding model space of Ricci curvature $-K$, then [13, Theorem 3.8] shows that

$$s \to \frac{I(x, s, \xi)}{I_K(s)}$$

is non-increasing. It follows that

$$J_{x,s}(y) = \frac{I(x, s, \xi)}{I(x, d(x,y), \xi)} \geq \frac{I_K(s)}{I_K(d(x,y))}$$

for all $s \in (0, d(x,y))$. Finally,

$$I_K(t) = \begin{cases} \left[\sqrt{\tfrac{n-1}{K}} \sinh\left(\sqrt{\tfrac{K}{n-1}}\, t \right) \right]^{n-1} & \text{if } K > 0 \\ t^{n-1} & \text{if } K = 0. \end{cases}$$

Thus, for $0 < s < t$,

$$\frac{I_K(s)}{I_K(t)} \geq \left(\frac{s}{t} \right)^{n-1} \exp\left(-\sqrt{(n-1)K}\, t \right).$$

Finally, we get

$$J_{x,s}(y) \geq \frac{1}{2^{n-1}} \exp\left(-\sqrt{(n-1)K}\, r \right)$$

for all $x \in M$, all y not in the cut locus of x and all $s \in [d(x,y)/2, d(x,y)]$. This proves Lemma 5.6.7.

5.7   Concluding remarks

In this last section we briefly indicate some further developments that emphasize some of the most basic features of the techniques presented in this monograph.

In Chapters 3, 4 and 5, we developed in the classical context of Riemannian manifolds a number of techniques based on Sobolev, Poincaré and other similar inequalities which allow us to study some of the fundamental properties of solutions of the heat equation

$$(\partial_t + \Delta)u = 0,$$

in particular Harnack-type inequalities. Beside Sobolev-type inequalities, we mostly based our analysis on a control of the volume growth of the manifold. In fact, we made no explicit use of the Riemannian structure. For instance, we did not place conditions on the curvature tensor except to show that some of the main results that we obtained do apply under certain curvature conditions.

It is indeed one of the advantages of the techniques presented in this monograph that they are immediately applicable outside the scope of Riemannian geometry, in particular in the context of "sub-Riemannian geometry". The simplest and most natural setting for an introduction to sub-Riemannian geometry is that of analysis on Lie groups. Let us identify the tangent space at the neutral element e of a Lie group G with the Lie algebra $\mathfrak{G}$ of G. Picking a (vector space) basis in $\mathfrak{G}$ amounts to picking a left-invariant Riemannian structure on G. However, from an algebraic point of view, it is natural to consider not necessarily a linear basis but a family of vectors $(X_1, \ldots, X_k)$ which generates $\mathfrak{G}$ as an algebra, i.e., $X_1, \ldots, X_k$ together with their brackets of all orders span the vector space $\mathfrak{G}$. This corresponds to the celebrated Hörmander subellipticity condition for the left-invariant differential operator $L = -\sum_1^k X_i^2$. See, e.g., [24, 44, 87].

To give an explicit example, consider the Heisenberg group $\mathbb{H}_1$ at (5.6.1). In this case, it is easy to see that the vector fields $X = \partial_x$ and $Y = \partial_y + x\partial_z$ generate the Lie algebra since the bracket $[X, Y]$ equals $Z = \partial_z$. Thus one is led to consider the subelliptic operator $L = -(X^2 + Y^2)$. This operator is, in many ways, more canonical than $-(X^2 + Y^2 + Z^2)$. For instance, L is homogeneous of degree two with respect to the natural dilation structure $(x, y, z) \mapsto (tx, ty, t^2 z)$. Using this, one easily shows that the coresponding volume growth function is $V(r) = cr^4$, $r > 0$. Analysis on Lie groups admitting a dilation structure is treated in [24].

Going back to a general Lie group, the techniques of this book (see in particular Theorem 5.6.1) easily yield a self-contained proof of the fact that the heat diffusion equation $(\partial_t + L)u = 0$ associated with a subelliptic operator L as above on a unimodular Lie group G has a continuous strictly positive fundamental solution $h(t, x, y)$ such that $h(t, x, x) \approx t^{-d/2}$ for small t where d is a certain integer that can be computed in terms of the family of left-invariant vector fields $\{X_1, \ldots, X_k\}$. All that is needed, in addition to what has been explained in this text, is control of the volume growth function in such a subelliptic situation. The courageous reader will find details and much more in [72, 87].

In fact, one natural setting for the development of the techniques presented in this text is that of a manifold M equipped with a measure μ and a second order differential operator L which is symmetric with respect to μ, that is such that $\int_M fLg d\mu = \int_M gLf d\mu$ for a large enough class of compactly supported functions. The length of the gradient of f can be defined in this context by setting

$$|\nabla f|^2 = -\frac{1}{2}(Lf^2 - 2fLf)$$

(assuming enough functions f in the domain of L are such that f^2 is also in the domain of L). At a formal level, one can also define the so-called intrinsic distance $d(x, y)$ associated to L by setting

$$d(x, y) = \sup\{f(x) - f(y) : f \text{ such that } |\nabla f| \leq 1\}.$$

Observe that, if L is the Laplace–Beltrami operator of a Riemannian manifold and μ is the Riemannian measure, then the intrinsic distance d is indeed the Riemannian distance. In general, whether or not the formula above really gives a genuine distance function which defines the topology of M is an interesting and deep question whose answer of course depends on certain assumptions made on L. See [25, 47, 74, 76, 87]. Even more generally, one can consider the geometry associated with strictly local regular Dirichlet forms. See [6, 51, 76, 83, 84] for pointers to the literature in this interesting direction.

Another important aspect of the methods used above is their great robustness. For instance, we proved the stability of the parabolic Harnack principle under quasi-isometry. This can be pushed further to treat stability under the so-called "rough isometries" which preserve the large scale geometry but not the local geometry or topology, a notion that has received great attention as it is central in some of the work and ideas of Gromov. See, e.g., [13, Section 4.4], [20] and the references therein. The simplest setting where this is useful is that of coverings of compact manifolds. Suppose N is a compact Riemannian manifold and M a Riemannian cover of N with deck transformation group Γ. This means that the finitely generated group Γ acts on M by isometries and $M/\Gamma = N$. A typical result that can be proved by using rough isometry techniques and the methods of this book is that, if Γ is a nilpotent group, then M satisfies the doubling condition and Poincaré inequality, uniformly at all scales. Thus, such a manifold satisfies the parabolic Harnack principle at all scales. It is interesting to note that this sort of result does not seem to be attainable by techniques based on curvature lower bounds.

Finally, it may be useful to recall that Moser's iteration technique applies to a host of other linear and quasi-linear equations. See, e.g., [4, 75, 76] and the references given there. In particular, it applies to the p-Laplacian $\operatorname{div}(|\nabla f|^{p-2}\nabla f)$ associated to the "energy functional" $\int_M |\nabla f|^p d\mu$. For instance, if a manifold M satisfies the doubling property and a scale-invariant L^p Poincaré inequality for some $p > 1$, then any non-negative p-harmonic function (i.e., solution of $\operatorname{div}(|\nabla f|^{p-2}\nabla u) = 0$) must be constant. This applies to Lie groups of polynomial volume growth, to manifolds with nonnegative Ricci curvature, and to coverings of compact manifolds with nilpotent deck transformation groups. When $p = n$ is the topological dimension of M the study of the p-Laplacian is relevant to the theory of quasiconformal (or quasi-regular) mappings. For developments and pointers to the literature in this direction see for instance [43, 42].

Bibliography

[1] Adams R. *Sobolev Spaces*. 1975, Academic Press.

[2] Aronson D.G. *Uniqueness of positive weak solutions of second order parabolic equations*. Annales Polonici Math. XVI, 1965, 286–303.

[3] Aronson D.G. *Bounds for the fundamental solution of a parabolic equation*. Bull. Amer. Math. Soc. 73, 1967, 890–896.

[4] Aronson D.G. and Serrin J. *Local behavior of solutions of quasilinear parabolic equations*. Arch. Ration. Mech. Anal. 25, 1967, 81–122.

[5] Aubin T. *Nonlinear Analysis on Manifolds. Monge–Ampère Equations*. 1982, Springer.

[6] Bakry D., Coulhon T., Ledoux M. and Saloff-Coste L. *Sobolev inequalities in disguise*. Indiana Univ. Math. J., 44, 1995, 1033–1073.

[7] Benjamini I., Chavel I. and Feldman E. *Heat kernel lower bounds on Riemannian manifolds using the old ideas of Nash*. Proc. London Math. Soc. 72, 1996, 215–240.

[8] Bojarski B. *Remarks on Sobolev imbedding inequalities* In *Complex Analysis*, Lect. Notes Math. 1351, 1989, Springer, 52–68.

[9] Bombieri E. and Giusti E. *Harnack's inequality for elliptic differential equations on minimal surfaces*. Invent. Math. 15, 1972, 24–46.

[10] Buser P. *A note on the isoperimetric constant*. Ann. Sci. École Norm. Sup. 15, 1982, 213–230.

[11] Carlen E., Kusuoka S. and Stroock D. *Upper bounds for symmetric Markov transition functions*. Ann. Inst. H. Poincaré, Prob. Stat., 23, 1987, 245–287.

[12] Chavel I. *Eigenvalues in Riemannian Geometry*. 1984, Academic Press.

[13] Chavel I. *Riemannian Geometry: A Modern Introduction*. 1993, Cambridge University Press.

[14] Cheeger J., Gromov M. and Taylor M. *Finite propagation speed, kernel estimates for functions of the Laplace operator and the geometry of complete manifolds.* J. Diff. Geom. 17, 1982, 15–23.

[15] Cheng S., Li P. and Yau S-T. *On the upper estimate of the heat kernel on a complete Riemannian manifold.* Amer. J. Math. 103, 1981, 1021–1036.

[16] Chua S. *Weighted Sobolev inequalities on domains satisfying the chain condition.* Proc. Amer. Math. Soc. 117, 1993, 449–457.

[17] Colding T. and Minicozzi II W. *Harmonic functions on manifolds.* Ann. Math. (2) 146, 1997, 725–747.

[18] Coulhon T. and Grigor'yan A. *On-diagonal lower bounds for heat kernels and Markov chains.* Duke Math. J., 89, 1997, 133–199.

[19] Coulhon T. and Saloff-Coste L. *Isopérimétrie pour les groupes et les variétés.* Revista Mat. Iberoamericana, 9, 1993, 293–314.

[20] Coulhon T. and Saloff-Coste L. *Variétés riemanniennes isométriques à l'infini.* Revista Mat. Iberoamericana, 11, 1995, 687–726.

[21] Davies E.B. *Heat Kernels and Spectral Theory.* 1989, Cambridge University Press.

[22] Davies E.B. *Non-Gaussian aspects of heat kernel behaviour.* J. London Math. Soc. 55, 1997, 105–125.

[23] Fabes, E. and Stroock D. *A new proof of Moser's parabolic Harnack inequality using the old ideas of nash.* Arch. Ration. Mech. Anal. 96, 1986, 327–338.

[24] Folland G. and Stein E. *Hardy Spaces on Homogeneous Groups.* 1982, Princeton University Press.

[25] Franchi B., Gutiérrez C. and Wheeden R. *Weighted Sobolev–Poincaré inequalities for Grushin-type operators.* Comm. Partial Differential Equations 19, 1994, 523-604.

[26] Fukushima M., Oshima Y. and Takeda M. *Dirichlet Forms and Symmetric Markov Processes.* 1994, Walter de Gruyter.

[27] Gaffney M.P. *The conservation property of the heat equation on Riemannian manifolds.* Comm. Pure Appl. Math. 12, 1959, 1–11.

[28] Gagliardo E. *Proprietà di alcune classi di funzioni in piu variabili.* Ric. Mat. 7, 1958, 102–137.

[29] Gallot S., Hulin D. and Lafontaine J. *Riemannian Geometry.* 2nd edn, 1990, Springer.

[30] Gilbarg D. and Trudinger N. *Elliptic Partial Differential Equations of Second Order.* 1977, Springer.

[31] Grigor'yan A. *On stochastically complete manifolds.* Soviet Math. Dokl., 34, 1987, 310–313.

[32] Grigor'yan A. *The heat equation on non-compact Riemannian manifolds.* Matem. Sbornik, 182, 1991, 55-87. Engl. Transl. Math. USSR Sb., 72, 1992, 47–77.

[33] Grigor'yan A. *Heat kernel upper bounds on a complete non-compact Riemannian manifold.* Revista Mat. Iberoamericana, 10, 1994, 395-452.

[34] Grigor'yan A. *Analytic and geometric background of recurrence and non-explosion of the Brownian motion on Riemannian manifolds.* Bull. Amer. Math. Soc. (N.S.), 36, 1999, 135–249.

[35] Grigor'yan A. and Saloff-Coste L. *Heat kernel on connected sums of Riemannian manifolds.* Mathematical Research Letters 6, 1999, 1–14.

[36] Guivarc'h Y. *Croissance polynômiale et périodes des fonctions harmoniques.* Bull. Soc. Math. France, 101, 1973, 333–379.

[37] Hadamard J. *Extention à l'équation de la chaleur d'un théorème de A. Harnack.* Rend. Circ. Mat. Palermo, Ser. 2, 3, 1954, 337–346.

[38] Hajlasz P. and Koskela P. *Sobolev met Poincaré.* Mem. Amer. Math. Soc. 145, 2000, no. 688.

[39] Hebey E. *Sobolev Spaces on Riemannian Manifolds.* Lect. Notes Math. 1635, 1996, Springer.

[40] Hebey E. *Nonlinear Analysis on Manifolds: Sobolev Spaces and Inequalities.* Courant Lecture Notes, AMS, 2000.

[41] Hebisch W. and Saloff-Coste L. *Gaussian estimates for Markov chains and random walks on groups.* Ann. Prob. 21, 1993, 673–709.

[42] Holopainen I., Coulhon T. and Saloff-Coste L. *Harnack inequality and hyperbolicity for subelliptic p-Laplacians with applications to Picard type theorems.* Geom. Funct. Anal., to appear.

[43] Holopainen I. and Rickman S. *Classification of Riemannian manifolds in nonlinear potential theory.* Potential Anal. 2, 1993, 37–66.

[44] Hörmander L. *Hypoelliptic second order differential equations.* Acta Math. 119, 1967, 147–171.

[45] Jenkins J. *Growth of connected locally compact groups.* J. Funct. Anal. 12, 1973, 113–127.

[46] Jerison D. *The Poincaré inequality for vector fields satisfying Hörmander's condition.* Duke Math. J. 53, 1986, 503–523.

[47] Jerison D. and Sanchez-Calle A. *Subelliptic second order differential operators.* In Complex Analysis III, Lect. Notes Math. 1277, 1986, Springer, 47–77.

[48] Knothe H. *Contribution to the theory of convex bodies.* Michigan Math. J. 1957, 39–52.

[49] Korevaar N. and Schoen R. *Global existence theorems for harmonic maps to non-locally compact spaces.* Comm. Anal. Geom. 5, 1997, 333–387.

[50] Kusuoka S. and Stroock D. *Application of Malliavin calculus, part 3.* J. Fac. Sci. Univ. Tokyo, Sect IA Math., 34, 1987, 391–442.

[51] Ledoux M. *The geometry of Markov diffusion operators,* Annales de la Faculté des Sciences de Toulouse, 9, 2000, 305–366.

[52] Levin D. and Solomyak M. *The Rozenblum–Lieb–Cwickel inequality for Markov generators.* J. Anal. Math. 71, 1997, 173–193.

[53] Li P. *On polynomial growth harmonic sections.* Math. Res. Lett. 4, 1997, 35–44.

[54] Li P. *Curvature and function theory on Riemannian manifolds.* Surveys in Diff. Geom., to appear.

[55] Li P. and Yau S-T. *On the Schrödinger equation and the eigenvalue problem.* Comm. Math. Phys. 88, 1983, 309–318.

[56] Lieb E. *Bounds on the eigenvalues of the Laplace and Schrödinger operators.* Bull. Amer. Math. Soc. 82, 1976, 751–753.

[57] Lu G. *The sharp Poincaré inequality for free vector fields: an endpoint result.* Revista Mat. Iberoamericana 10, 1994, 453–466.

[58] Lyons T. *Instability of the Liouville property for quasi-isometric Riemannian manifolds and reversible Markov chains.* J. Diff. Geom. 19, 1987, 33–66.

[59] Maheux P. *Analyse sur les nil-variétés.* Thèse, University Paris VI.

[60] Maz'ja W. *Sobolev Spaces.* 1985, Springer.

[61] Maz'ja W. and Poborchi S.V. *Differentiable Functions on Bad Domains.* 1997, World Scientific.

[62] Milman V. and Schechtman G. *Asymptotic Theory of Finite Dimensional Normed Spaces.* Lect. Notes Math. 1200, 1986, Springer.

[63] Moser J. *On Harnack's theorem for elliptic differential equations.* Comm. Pure Appl. Math. 14, 1961, 577–591.

[64] Moser J. *A Harnack inequality for parabolic differential equations.* Comm. Pure Appl. Math. 16, 1964, 101–134. Correction in 20, 1967, 231–236.

[65] Moser J. *On a pointwise estimate for parabolic differential equations.* Comm. Pure Appl. Math. 24, 1971, 727–740.

[66] Moser J. *A sharp form of an inequality by N. Trudinger.* Indiana Univ. Math. J. 20, 1971, 1077–1092.

[67] Nash J. *Continuity of solutions of parabolic and elliptic equations.* Amer. J. Math. 80, 1958, 931–954.

[68] Nirenberg L. *On elliptic differential equations.* Scuola Norm. Sup. Pisa, Sci. Fis. Mat. 13, 1959, 116–162.

[69] Pazy A. *Semigroups of Linear Operators and Applications to Partial Differential Equations.* Springer, 1983.

[70] Pini B., *Sulla soluzione generalizzata di Wiener per il primo problema di valori al contorno nel caso parabolico.* Rend. Sem. Mat. Univ. Padova, 23, 1954, 422–434.

[71] Porper F. and Eidel'man S. *Two-sided estimates of fundamental solutions of second-order parabolic equations and some applications.* Russian Math. Surveys 39, 1984, 119–178.

[72] Robinson D. *Elliptic Operators and Lie Groups.* Oxford Universiy Press, 1991.

[73] Rozenblyum G. and Solomyak M. *The Cwikel–Lieb–Rozenblyum estimator for generators of positive semigroups and semigroups dominated by positive semigroups.* St. Petersburg Math. J. 9, 1998, 1195–1211.

[74] Saloff-Coste L. *A note on Poincaré, Sobolev and Harnack inequalities.* Duke Math. J., 65, IMRN, 2, 1992, 27–38.

[75] Saloff-Coste L. *Uniformly elliptic operators on Riemannian manifolds.* J. Diff. Geom. 36, 1992, 417–450.

[76] Saloff-Coste L. *Parabolic Harnack inequality for divergence form second order differential operators.* Potential Analysis 4, 1995, 429–467.

[77] Saloff-Coste L. and Stroock D. *Opérateurs uniformément sous-elliptiques sur les groupes de Lie.* J. Funct. Anal. 98, 1991, 97–121.

[78] Sobolev S.L. *On a theorem of functional analysis.* Mat. Sb. (N.S.) 4, 1938, 471-479, English transl., AMS Transl. Ser 2, 34, 1963, 36-68.

[79] Stein E. *Singular Integrals and Differentiability Properties of Functions.* 1970, Princeton University Press.

[80] Stein E. *Topics in Harmonic Analysis Related to the Littlewood-Paley Theory.* Annals of Mathematical Studies 63, 1970, Princeton University Press.

[81] Stein E. *Harmonic Analysis.* 1993, Princeton University Press.

[82] Stein E. and Weiss G. *Introduction to Fourier Analysis on Euclidean Spaces.* 1971, Princeton University Press.

[83] Sturm K. *On the geometry defined by Dirichlet forms.* In "Seminar on Stochastic Analysis, Random Fields and Applications" (Ascona 1993, E. Bolthausen *et al.*, eds.), 1995, 231–242, Birkhäuser.

[84] Sturm K. *The geometric aspect of Dirichlet forms.* In "New directions in Dirichlet forms" AMS/IP Stud. Adv. Math. 8, 1998, 233–277.

[85] Talenti G. *Best constant in Sobolev inequality.* Ann. Mat. Pura Appl. 110, 1976, 353–372.

[86] Trudinger N. *On imbedding into Orlitz spaces and some applications.* J. Math. Phys. 17, 1967, 473–484.

[87] Varopoulos N., Saloff-Coste L. and Coulhon T. *Geometry and Analysis on Groups.* 1993, Cambridge University Press.

[88] Yau S-T. *Harmonic functions on complete Riemannian manifolds.* Comm. Pure Appl. Math. 28, 1975, 201–228.

[89] Yau S-T. *Survey on partial differential equations in differential geometry* In *Seminar in differential geometry*, Annals of Mathematical Studies, 102, 3–71, 1982, Princeton University Press.

[90] Yudovich V. *On certain estimates connected with integral operators and solutions of elliptic equations.* Dokl. Akad. Nauk SSSR 138:4, 1961, 805-808 (English translation: Soviet Math., Vol. 2, 3, 1961, 746-749).

Index

Aronson (Donald) 6, 156

Birman–Schwinger principle 109

Bishop's theorem 83

Bishop-Gromov theorem 84, 179

Bombieri (Enrico) 47

Brownian motion 161

Brunn (Hermann) 18

Brun-Minkowski inequality 18

co-area formula 17, 21, 56, 76

covering (Whitney type) 133, 135, 138, 141

covering manifold 182

deck transformation group 182

De Giorgi (Ennio) 50

dilation structure 181

Dirichlet
 boundary condition 29
 eigenvalue 167
 eigenvalue problem 29
 form 87, 106, 182
 heat kernel 161, 166
 semigroup 166

divergence 19, 54

divergence form 2, 33, 49, 68

doubling property 5, 112, 117, 119, 127, 139, 154, 155, 173, 175

Faber–Krahn inequality 164

Federer (Herbert) 17

Fleming (Wendell) 17

Gagliardo (Emilio) 9, 17

Gaussian estimate 93
 lower 127, 131
 two-sided 6, 154, 161, 184
 upper 4, 99, 101, 122, 165

Giusti (Enrico) 47

gradient 7, 11, 54, 175, 181

Green function 154

Gromov (Misha) 18, 182

Haar measure 77, 172

harmonic function 111, 151

Harnack inequality(principle) 145, 146
 elliptic 2, 35, 49, 111, 151
 gradient 113
 parabolic 5, 111, 155, 164, 168, 175

heat diffusion semigroup 4, 87, 95, 98, 122

heat equation 4, 112, 119, 127, 146

heat kernel 89, 93, 101, 122, 128,
 estimate(s) 92, 154
 on-diagonal lower bound 125, 127, 152, 165
 upper bound 3, 130

Heisenberg group 82, 173

Hölder continuity estimate 35, 50, 149

Hölder inequality 9, 45, 57, 60, 68, 66

Hörmander condition 181

intrinsic distance 181

isoperimetric inequality 16, 21, 56, 81
 problem 55

John–Nirenberg inequality 47

Knothe (Herbert) 18

Laplace–Beltrami operator 54, 78, 103, 161, 175, 176